Plant-Soil Interactions

Plant-Soil Interactions

Editor

Michel-Pierre Faucon

MDPI • Basel • Beijing • Wuhan • Barcelona • Belgrade • Manchester • Tokyo • Cluj • Tianjin

Editor
Michel-Pierre Faucon
Institut Polytechnique
UniLaSalle
France

Editorial Office
MDPI
St. Alban-Anlage 66
4052 Basel, Switzerland

This is a reprint of articles from the Special Issue published online in the open access journal *Diversity* (ISSN 1424-2818) (available at: https://www.mdpi.com/journal/diversity/special_issues/plant_soil_interactions).

For citation purposes, cite each article independently as indicated on the article page online and as indicated below:

LastName, A.A.; LastName, B.B.; LastName, C.C. Article Title. *Journal Name* **Year**, *Volume Number*, Page Range.

ISBN 978-3-0365-0406-3 (Hbk)
ISBN 978-3-0365-0407-0 (PDF)

Cover image courtesy of Sophie Faucon Potier.

Contents

About the Editor

Michel-Pierre Faucon was born on 5 October 1982 in Bethune (France, 62) and is Ass. Professor of plant ecology and agroecology in the Institut Polytechnique UniLaSalle (ULS) where he is Director of Research and Innovation. He lectures in plant ecology, functional ecology, agroecology and ecological engineering. He holds a PhD in plant ecology and biogeochemistry (2005–2009, Free University of Brussels) and accreditation to supervise research and PhDs (2015). His research focuses on functional ecology of plant–soil interactions in grassland ecosystems and agroecosystems. He has developed trait-based agroecology in new cropping systems including multispecific crops for food, bio-sourced energy and ore to understand plant–soil feedback processes, manage ecosystem services and develop bioeconomy. Recent research has allowed highlighting the key role of the functional structure of cover crops in phosphorus availability and soil physical properties involved in soil resistance to erosion and traffic-induced soil deformation. He has acted as project leader in more than ten funded research projects (funds from European Union, French Ministry of Research and Education, private companies in environment and agriculture). He has published more than 50 peer-reviewed papers and frequently acts as chairman in international conferences. He has successfully supervised five PhD students. He is a member of several professional organizations and coleader of innovation projects for sustainable and inclusive territorial development.

Editorial

Plant–Soil Interactions as Drivers of the Structure and Functions of Plant Communities

Michel-Pierre Faucon

AGHYLE (SFR Condorcet FR CNRS 3417), UniLaSalle, 19 Rue Pierre Waguet, 60026 Beauvais, France; michel-pierre.faucon@unilasalle.fr; Tel.: +33-(0)3-44-06-38-03

Received: 23 November 2020; Accepted: 27 November 2020; Published: 28 November 2020

1. Introduction

Plant–soil interactions play an important role in the structure and function of plant communities and thus in the functioning of ecosystems. Soil properties represent a strong selection pressure for plant diversity and influence the structure of plant communities and participate in the generation and maintenance of biodiversity. Plant communities grow by modifying the physical, chemical, and biological properties of soil, with consequent effects on survival and growth of plants. This process—called plant–soil feedback—plays a key role in water and nutrient availability and the dynamic of soil-borne microbial pathogens, parasite populations, and root herbivores, by globally impacting the vegetation succession and net primary productivity. Plant–soil feedbacks can, however, present contrasting effects on plant community assemblage and ecosystem development [1].

Plant–soil interactions include many biophysical and chemical processes combined with spatial and temporal variations and represent an ecological complexity that still needs to be unraveled. The challenge is to improve knowledge on the role of plant–soil interactions in the plant community structure and functions and their consequences in ecosystem functioning. Functional trait approaches enhance ecological process understanding by characterizing the mechanisms that govern interactions between plants and their environments, and, particularly in this Special Issue, soil–plant interactions [2]. This "plant–soil interactions" Special Issue addresses both soil factor effects on plant communities and the role of ecological complementarity and species diversity of plant communities in soil properties and ecosystem services.

Soil factor effects on plant community structure are addressed by highlighting plant trait selection by degraded habitats in ultramafic soils for mine reclamation [3], by specifying the effect of microorganism supply (AMF, *Frankia* and *Rhizophagus fasciculatus*) on plant community structures and ecological restoration [4], and by reviewing the role of Arbuscular Mycorrhizal Fungi (AMF) to cope with biotic and abiotic stresses and improve plant growth [5]. In addition, a distinct composition of plant-associated core bacterial communities independent of land use intensity is identified [6].

The role of ecological complementarity and species diversity of plant communities in soil properties and ecosystem services is addressed in this Special Issue. Ecological complementarity is demonstrated by the integration of legumes in nitrogen deficiency monospecific tropical grass pasture increasing biomass productivity and N uptake [7].

The non-additive effect of functional diversity (FD) is demonstrated on several soil processes and ecosystem services. The non-additive effect of FD on the hydraulic resistance and thus soil sedimentation is shown and can be explained by the presence of species presenting large stems in the communities with high functional diversity [8]. The effect of plant traits on hydraulic resistance and soil sedimentation process is driven by the community-weighted leaf density and not by the functional diversity of leaf and stem traits at the community level [8]. In another ecological process, species diversity does not show a positive effect on the abundance of fungal pathogens. White mustard is the only cover crop species associated with the decrease in necrotic root damage and abundance of fungal pathogens in vineyards [9].

This plant–soil interactions Special Issue brings new empirical results and some advances in order to try to unravel the complexity of plant–soil interactions and their role in plant community structure and function.

Funding: This research received no external funding.

Conflicts of Interest: The author declares no conflict of interest.

References

1. Kardol, P.; Martijn Bezemer, T.; van der Putten, W. Temporal variation in plant–soil feedback controls succession. *Ecol. Lett.* **2006**, *9*, 1080–1088. [CrossRef] [PubMed]
2. Faucon, M.-P.; Houben, D.; Lambers, H. Plant functional traits: Soil and ecosystem services. *Trends Plant Sci.* **2017**, *22*, 385–394. [CrossRef] [PubMed]
3. Quintela-Sabarís, C.; Faucon, M.-P.; Repin, R.; Sugau, J.; Nilus, R.; Echevarria, G.; Leguédois, S. Plant Functional Traits on Tropical Ultramafic Habitats Affected by Fire and Mining: Insights for Reclamation. *Diversity* **2020**, *12*, 248. [CrossRef]
4. Djighaly, P.; Ngom, D.; Diagne, N.; Fall, D.; Ngom, M.; Diouf, D.; Hocher, V.; Laplaze, L.; Champion, A.; Farrant, J.; et al. Effect of *Casuarina* Plantations Inoculated with Arbuscular Mycorrhizal Fungi and *Frankia* on the Diversity of Herbaceous Vegetation in Saline Environments in Senegal. *Diversity* **2020**, *12*, 293. [CrossRef]
5. Diagne, N.; Ngom, M.; Djighaly, P.; Fall, D.; Hocher, V.; Svistoonoff, S. Roles of Arbuscular Mycorrhizal Fungi on Plant Growth and Performance: Importance in Biotic and Abiotic Stressed Regulation. *Diversity* **2020**, *12*, 370. [CrossRef]
6. Estendorfer, J.; Stempfhuber, B.; Vestergaard, G.; Schulz, S.; Rillig, M.; Joshi, J.; Schröder, P.; Schloter, M. Definition of Core Bacterial Taxa in Different Root Compartments of Dactylis glomerata, Grown in Soil under Different Levels of Land Use Intensity. *Diversity* **2020**, *12*, 392. [CrossRef]
7. Villegas, D.; Velasquez, J.; Arango, J.; Obregon, K.; Rao, I.; Rosas, G.; Oberson, A. Urochloa Grasses Swap Nitrogen Source When Grown in Association with Legumes in Tropical Pastures. *Diversity* **2020**, *12*, 419. [CrossRef]
8. Kervroëdan, L.; Armand, R.; Saunier, M.; Faucon, M.-P. Functional Diversity Effects of Vegetation on Runoff to Design Herbaceous Hedges for Sediment Retention. *Diversity* **2020**, *12*, 131. [CrossRef]
9. Richards, A.; Estaki, M.; Úrbez-Torres, J.; Bowen, P.; Lowery, T.; Hart, M. Cover Crop Diversity as a Tool to Mitigate Vine Decline and Reduce Pathogens in Vineyard Soils. *Diversity* **2020**, *12*, 128. [CrossRef]

Article

Plant Functional Traits on Tropical Ultramafic Habitats Affected by Fire and Mining: Insights for Reclamation

Celestino Quintela-Sabarís [1,*], Michel-Pierre Faucon [2], Rimi Repin [3], John B. Sugau [4], Reuben Nilus [4], Guillaume Echevarria [1,5] and Sophie Leguédois [1]

[1]   INRAE, Université de Lorraine, LSE, F-54000 Nancy, France; Guillaume.Echevarria@univ-lorraine.fr (G.E.); sophie.leguedois@univ-lorraine.fr (S.L.)
[2]   AGHYLE, UP 2018.C101, SFR Condorcet FR CNRS 3417, UniLaSalle, 60026 Beauvais, France; michel-pierre.faucon@unilasalle.fr
[3]   Research and Education Division, Sabah Parks, Kota Kinabalu 88806, Sabah, Malaysia; sparks.researchedu@gmail.com
[4]   Forest Research Centre, Sabah Forestry Department, Sandakan 90715, Sabah, Malaysia; john.sugau@sabah.gov.my (J.B.S.); reuben.nilus@sabah.gov.my (R.N.)
[5]   Centre for Mined Land Rehabilitation, SMI, University of Queensland, St. Lucia QLD 4072, Australia
*   Correspondence: tino.quintela.sabaris@gmail.com

Received: 24 April 2020; Accepted: 5 June 2020; Published: 17 June 2020

Abstract: Biodiversity-rich tropical ultramafic areas are currently being impacted by land clearing and particularly by mine activities. The reclamation of ultramafic degraded areas requires a knowledge of pioneer plant species. The objective of this study is to highlight the functional traits of plants that colonize ultramafic areas after disturbance by fire or mining activities. This information will allow trait-assisted selection of candidate species for reclamation. Fifteen plots were established on ultramafic soils in Sabah (Borneo, Malaysia) disturbed by recurrent fires (FIRE plots) or by soil excavation and quarrying (MINE plots). In each plot, soil samples were collected and plant cover as well as species abundances were estimated. Fifteen functional traits related to revegetation, nutrient improvement, or Ni phytomining were measured in sampled plants. Vegetation of both FIRE and MINE plots was dominated by perennials with lateral spreading capacity (mainly by rhizomes). Plant communities displayed a conservative growth strategy, which is an adaptation to low nutrient availability on ultramafic soils. Plant height was higher in FIRE than in MINE plots, whereas the number of stems per plant was higher in MINE plots. Perennial plants with lateral spreading capacity and a conservative growth strategy would be the first choice for the reclamation of ultramafic degraded areas. Additional notes for increasing nutrient cycling, managing competition, and implementing of Ni-phytomining are also provided.

Keywords: community weighted means; functional traits; soil reclamation; technosols; ultramafic

---

1. Introduction

Ultramafic soils are ecological or 'edaphic islands' due to their patchy distribution and contrasting soil conditions with respect to surrounding 'normal' soils [1,2]. Several extreme soil factors including macronutrient deficiency (N, P, K, Ca), Mg toxicity resulting in extremely low Ca:Mg molar ratio, and highly plant-available trace elements (Ni, Cr, Co) make ultramafic areas a stressful environment for plant establishment and growth [3,4]. The extreme edaphic conditions and isolated island-like distribution of ultramafic soils has led to the origin of numerous strict ultramafic endemic plant species, particularly in tropical regions, such as Cuba, New Caledonia, and Southeast Asia [5–7]. Ultramafic areas in Sabah (North of Borneo, Malaysia) support a rich flora with more than 4500 species described to date and a very high proportion of strict endemics [8]. Besides its taxonomic and evolutionary interest [9], ultramafic flora is a remarkable biological resource for eco-technological applications, especially phytoremediation of contaminated soils [10].

Ultramafic areas have been extensively mined for the recovery of different metals such as Ni (Ni sulphide deposits and Ni laterites) and Cr (chromite) [11]. As the few high-grade Ni sulphide deposits have become depleted, mining for Ni has shifted focus to Ni laterite deposits in tropical areas including Australia, Cuba, New Caledonia, Brazil, and Indonesia [12]. In comparison to localized open pit mining of Ni sulphide, Ni laterite mining is highly destructive to ecosystems since it involves complete removal of vegetation and topsoil over a large area (strip mining) to access the Ni-rich saprolite below and some of the laterite [11,13]. Removal of the topsoil limits nutrient and water buffering capacity vital to the development of vegetation [11,14]. Logging and land clearing (mainly using fire) is another threat to ultramafic ecosystems, especially in Southeast Asia [7,15,16]. Logging and wildfires are less destructive than mining because they affect primarily the vegetation, leaving the soil more or less intact. However, after fires significant soil erosion and loss of carbon and some nutrients may occur. Tropical forest ecosystems may take more than a decade to fully recover after logging, whereas for mined areas it may take up to 250 years as the soil regenerates [17,18].

Pioneer plant communities on disturbed areas are derived from the local species pool and the effect of environmental filters (either stringent soil conditions or biotic interactions) [19]. It has been shown that experimental plant communities with different species composition subjected to similar environmental conditions experience a convergence in plant traits [20]. That is, successful plant species are those possessing the best traits that convey tolerance to the specific environmental stressors.

Plant functional traits are morpho-physio-phenological traits which affect fitness indirectly via their effects on growth, reproduction, and survival [19,21]. Trait-based ecology associated with trait data measured across many individuals and species can be used to predict emergent properties of communities and ecosystems. The functional trait approach allows for the characterization of plant responses to the environment [22] and their effects on ecosystem function and services such as nutrient availability or soil carbon storage [23,24].

Thus, the study of functional traits of plant communities that spontaneously colonize disturbed ultramafic areas may provide information for trait-assisted selection of candidate species for reclamation of tropical ultramafic degraded areas. An additional benefit of a functional approach is that the obtained information on traits can be transferred to other sites with similar environmental conditions, without the local/regional species pool limits that affect floristic approaches [25]. This approach has been previously applied to the revegetation of copper-cobalt mine areas [25], to the restoration of quarries [26], or to predict the colonization of post-mining sites during spontaneous revegetation [27]. Moreover, the combination of plant species with complementary traits has shown benefits for the phytoremediation of polluted soils and the restoration of mine tailings [28,29].

In order to examine the effects of disturbance severity on resulting vegetation type on tropical ultramafic areas, we studied the soil properties and plant communities of areas that experienced moderate disturbance (wildfire; FIRE) and severe disturbance (mining; MINE). Community weighted mean (CWM) represents the most probable value for a certain trait in a plant randomly sampled from a community [19]. Different studies based on CWMs have found changes in functional traits in relation to environmental factors, either on Mediterranean abandoned vineyards [21] or in tropical dry and tropical wet forests [30]. Here, CWMs were calculated to characterize functional response of communities according to type of habitat degradation and soil factors, and compared to the functional traits of general vegetation [31] and vegetation from non-disturbed ultramafic areas from Sabah [32].

Study goals included: (i) to examine how the type of disturbance (wildfire and mining) affects soil parameters and CWM of traits of pioneer plant communities on ultramafic soils, and (ii) to identify important traits for the trait-assisted selection of plant species for the reclamation of tropical degraded ultramafic areas.

2. Materials and Methods

2.1. Study Area

Ultramafic soils in Sabah (north of Borneo, Malaysia) cover around 3500 km^2 [33], with the more extensive ultramafic outcrops found around Mount Kinabalu, Morou Porou, Bidu-Bidu Hills, Meliau Range, Mount Tawai, and Mount Silam [8]. Our research sites were located on three ultramafic degraded areas southeast of Mount Kinabalu, including Garas-Lompoyou hill chain, Bukit Hampuan Forest Reserve, and Paliu area. These areas contain sites degraded by wildfire (FIRE) or quarrying (MINE) (Figure 1). Fires in Sabah are linked to El Niño Southern Oscillation (ENSO) drought events and affect mainly logged forest, which are more prone to fire than undisturbed forests [34]. Most of the FIRE sites were logged forests affected by severe fires which affected Sabah during the 1997/98 ENSO. MINE sites are the result of quarrying serpentinite bedrock for road base aggregate (abandoned in 1999) or rocky dumpsites (downhill sidecast) created during road construction in 2010. The four plots in Bukit Hampuan FR were in the vicinity of primary cloud forest on ultramafic substrate, whereas the plots in Garas-Lompoyou and Paliu were in a mosaic landscape composed by cultivated land and secondary vegetation in different degrees of succession (from fern-dominated areas to secondary forests).

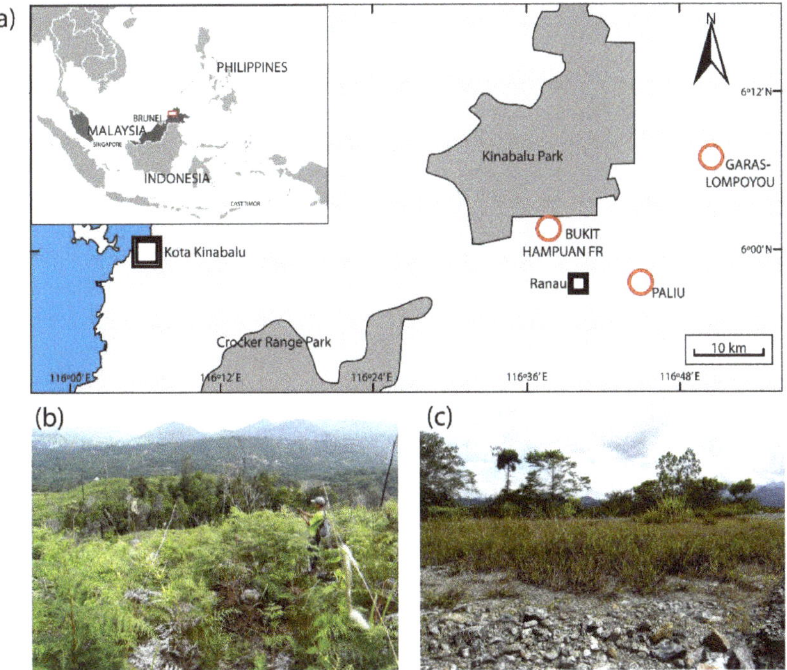

Figure 1. Overview of the studied area. (**a**) Position of sampling sites (marked as red circles). Main cities in the region are marked (with squares) for reference. Dark grey areas indicate Natural Parks. Position of the sampling area (red rectangle) in the context of South-East Asia is presented in the top-left insert. (**b**) General view of the vegetation in a FIRE plot. (**c**) General view of the vegetation in a MINE plot.

The climate of Sabah is tropical (Köppen climate Af). The mean annual temperature is 23 °C with low variation (less than 3 °C) throughout the year. Annual rainfall is around 2500 mm with relatively even rainfall throughout the year. Two notably less humid periods occur in February and August (see [35,36] and Malaysian meteorological department).

2.2. Plant and Soil Sampling

During July 2016, 15 circular non-permanent plots of 10 m radius were established on serpentinite-derived soils affected by soil excavation (MINE sites, eight plots) or by logging and wildfires (FIRE sites, seven plots). FIRE plots were in steep slopes (average 44%), whereas MINE plots were more flat (average slope 12%). Most of the plots had a south aspect. Plots in Paliu and Garas-Lompoyou area were in altitudes from 336 to 464 m asl, whereas the plots in Bukit Hampuan FR were around 1200 m asl. Soils in MINE sites were Spolic Technosols, whereas in FIRE sites soils were Cambic Leptosols [37,38]. In each plot, three radial transects separated by 120° were randomly established using a table of random numbers and a compass. In FIRE plots plant cover was estimated by line-intercept method, whereas in MINE plots vegetation cover was estimated by point-intercept method (one pin each 25 cm). This difference in methods was caused by the difference of vegetation height that made the use of the point-intercept method in FIRE plots not feasible. Both methods allowed the determination of percentage of cover for each species, as well as the percentage of bare soil.

In order to describe soil conditions, one representative soil sample (0–10 cm) was collected from the middle point of each transect (i.e., three soil samples per plot).

2.3. Soil Analyses

Fresh soil samples were sieved upon sampling and the >5 mm fraction was kept in plastic ziplock bags and stored for 8–10 weeks at 4 °C until analyses. The activity of four soil enzymes linked to the cycles of phosphorus (alkaline phosphatase), sulphur (arylsulphatase), carbon (β-glucosidase), and nitrogen (urease), and the hydrolysis of fluorescein diacetate (FDA, considered a proxy for the hydrolytic activity of the soil) were determined in fresh soil subsamples [39,40].

Soil enzyme activities were determined using colorimetric methods as indicated in [41]. The values of soil enzyme activities were expressed on an oven-dried soil basis.

Soil subsamples were air-dried and sieved at 2 mm. Water retention data were determined on a pressure plate apparatus for two water potentials (−10 and −15,800 kPa) [42]. Available water storage (AWS; g 100 g^{-1}), i.e., water disposable for plant growth was calculated by the following equation: AWS = (Wfc − Wwp), (1), where Wfc is the water content at field capacity (water potential: −10 kPa) (g 100 g^{-1}) and Wwp is the water content at permanent wilting point (water potential: −15,800 kPa) (g 100 g^{-1}).

Soil pH was measured in H_2O using a 1:5 (v/v) ratio. Cation exchange capacity (CEC) was determined colorimetrically after treatment of the soil with a solution of cobaltihexamine trichloride 0.05 N [43]. The filtered soil:cobaltihexamine extracts were analyzed by means of Inductively Coupled Plasma-Atomic Emission Spectrometry (ICP-AES, Liberty II, Varian Inc, Australia) to determine the soil exchangeable concentrations of Ca^{+2}, Mg^{+2}, and K^+. Soil available phosphorus (Olsen-P) was extracted with a solution of $NaHCO_3$ and quantified by reaction with ascorbic acid [44]. Soil nickel availability was evaluated after extraction with DTPA-TEA at pH 7.3, 1:2 w/v, 2 h shaking) [45]. Soil subsamples were ground in a ceramic mortar. Total soil C and N was estimated by combustion in a CHNS analyzer (Vario Micro Cube, Elementar, Germany). Dry ground soil subsamples (0.5 g) were digested in 2 mL of concentrated HNO_3 and 6 mL of concentrated HCl on a hot plate at 105 °C. Final solutions were filtered (0.45 μm DigiFILTER, SCP science, Canada) and diluted to 50 mL with deionized water. Pseudo-total soil concentrations of Co, Cr, Mn, Ni, P, and S were estimated by ICP-AES (Liberty II, Varian).

2.4. Plant Analyses

Individuals from each species identified in each plot were sampled for the determination of 15 functional traits related to plant persistence, nutrient management, and tolerance to ultramafic soils. We chose these traits because of their interest in different aspects of reclamation of disturbed ultramafic areas: revegetation, limitation of erosion, implementation of phytomining, and soil nutrient improvement. Besides usual functional traits (such as specific leaf area—SLA), we included six elemental concentration traits important to explain plant response to the particular conditions of ultramafic soils: Ni hyperaccumulation (Ni > 1000 μg g^{-1}); leaf tissue elemental concentrations of K, Ca, Mg, and Mn; and the Ca:Mg molar ratio of leaves (Table 1). Whole-plant traits (such as lateral spreading capacity or plant height) were assessed directly on the field in at least three plants per species. One plant per species was excavated to estimate rooting depth. Moreover, branches or shoots of 1–3 plants per species were collected, kept in sealed plastic bags, and transported to Monggis Substation (Kinabalu Park), where they were processed the day of the sampling. One healthy leaf per plant was selected and put below a glass layer and photographed with a digital camera. The camera was placed in a fixed support to guarantee its orthogonal position with respect to leaves. In some cases, the leaf was cut in several fragments to ensure the correct estimation of leaf area. All leaf photographs included a ruler to allow the estimation of leaf area. Leaf area was obtained from the digital photographs using ImageJ software [46]. After photographing, each leaf was cleaned with tap water, rinsed with deionized water, and put in a paper envelope. Leaf samples were kept in an oven (60 °C) for several weeks and dry leaf mass was obtained. Specific leaf area (SLA) was calculated as the ratio between leaf area and dry leaf mass, including petiole [47]. After dry weight was recorded, dry leaves were finely ground using a ball mill. Subsamples (0.5 g) of dry and ground tissue were digested at 95 °C in 2.5 mL of concentrated HNO$_3$ and 5 mL of H$_2$O$_2$ (30%). The final solutions were filtered (0.45 μm DigiFILTER) and diluted to 25 mL with deionized water. Leaf P, K, Ca, Mg, Mn, and Ni concentrations were measured by ICP-AES (Liberty II, Varian). Leaf C and N were quantified in dry ground leaves using a CHNS analyzer (Vario Micro Cube, Elementar, Germany).

2.5. Data Analysis

The soil dataset, including soil pH, soil water retention capacity (AWS, Wfc, Wwp), P-Olsen, total soil C and N, soil CEC, soil plant-available (exchangeable) Ca, K, and Mg, Ca:Mg molar ratio, DTPA-extractable Ni, pseudototal concentrations of P, S, Mn, Ni, Cr, and Co, and enzyme activities was analyzed using principal component analysis (PCA). PCA was based on a correlation matrix in order to account for differences in metrics among variables and no further standardization was applied. Differences in soil variables between type of degraded sites (FIRE vs. MINE) were further assessed by nested ANOVA analyses, with factors 'Type' and 'Site' nested within 'Type'. Plant species cover in each plot was used to compute Shannon's H diversity index. Differences in plant communities between FIRE and MINE sites were further assessed by non-metric multidimensional scaling (NMDS) using Bray–Curtis distances (hereafter referred as taxonomic NMDS), and constraining solution to only two dimensions. Shannon's index and NMDS were computed with functions *diversity* and *metaMDS* from the package *vegan* (ver 2.5-1) for R (ver. 3.4.4) [48].

A community weighted mean (CWM) was computed for each functional trait and for each plot applying the following formula: CWM = Σ (p_i × *trait$_i$*), (2), where p_i is the relative contribution of species *i* to the total plant cover of the community and *trait$_i$* is the trait value of species *i* [21]. In the case of binary variables CWM indicates the frequency of the presence of a trait. In the case of ordinal variables, we kept the most common value in each plot. CWM has shown to be a reliable parameter that is not affected by differences in methods for the estimation of plant relative abundance or the trait values [49]. CWMs were computed using the functions *functcomp* from the package FD (ver 1.0-12) for R [50].

The CWM data on the 15 studied plots were used to perform a second NMDS analysis (hereafter referred as functional NMDS) using Gower distance with the function *metaMDS* from *vegan* package [48]. Solution was constrained to only two dimensions. Environmental factors (soil parameters, altitude, slope, aspect, time since disturbance) were fitted as vectors onto the taxonomic and functional NMDS ordinations using the function *envfit* from the package *vegan* [48]. The correlation of the environmental vectors with the NMDS ordination and the *p*-value of that correlation were estimated by 1000 permutations. Only the environmental factors with *p*-values < 0.05 were plotted onto the NMDS graphs.

Differences in taxonomic diversity (Shannon's Index) between MINE and FIRE sites were compared by one-way ANOVAs. Non-parametrical Mann–Whitney U tests were applied to compare CWM of binary and ordinal traits between MINE and FIRE plots, whereas one-way ANOVAs were used for the comparison of CWM of quantitative traits. When necessary, data were log-transformed to meet ANOVA assumptions. PCA, ANOVAs, and Mann–Whitney U tests were computed using SPSS (v. 15, SPSS Inc., Chicago, IL, USA).

Table 1. List of the plant functional traits assessed in the sampled species. For each trait we include the unit, the categories (for traits coded as binary or as ordinal variables), the associated ecological functions, and the interests for reclamation of degraded ultramafic habitats.

Trait	Units	Categories/Domain	Associated Ecological Functions	Interest for Reclamation
Life cycle	Unitless	(0) annual (1) perennial	Response to disturbance and soil resources, competitive strength	
Lateral spreading capacity	Unitless	(0) absence (1) presence	Competitive strength	
Depth of root system	In cm	(1) 0–10 (2) 10–30 (3) >30	Response to disturbance and soil resources, competitive strength	Revegetation and/or limitation of erosion
Plant height	In m	(1) 0–0.11 (2) 0.11–0.29 (3) 0.30–0.59 (4) 0.60–0.99 (5) 1–3 (6) >3 m	Response to disturbance and soil resources, competitive strength	
Density of stems	Number of stems in 1 dm^2	(1) 1–10 (2) 10–30 (3) >30)	Competitive strength	
Specific leaf area (SLA)	$mm^2\,mg^{-1}$	Positive decimal value	Response to soil resources, plant defense	
N_2 fixation	Unitless	(0) absence (1), presence	Response to soil resources, nutrient strategy	
Leaf N concentration (LNC)	$mg\,g^{-1}$	Positive decimal value	Response to soil resources, influence in nutrient cycling	Soil nutrient improvement
Leaf P concentration (LPC)	$mg\,g^{-1}$	Positive decimal value	Response to soil resources, influence in nutrient cycling	
Leaf concentrations of Ca, Mg, K and Mn	$mg\,g^{-1}$	Positive decimal value	Nutrient strategy/response to ultramafic conditions	
Leaf Ca/Mg ratio	Unitless	Positive decimal value	Nutrient strategy/response to ultramafic conditions	
Ni hyperaccumulation	Unitless	(0) absence (1) presence	Response to ultramafic conditions	Phytomining

3. Results

3.1. Soil Parameters

Observed values of soil variables are typical for ultramafic soils with pH around 7 (neutral); Ca:Mg molar ratio <1; low concentrations of important nutrients (e.g., P-Olsen concentrations lower than 3 mg kg^{-1}); high pseudototal concentrations of Ni, Cr, and Co (mean values around 2400, 3500, and 275 mg kg^{-1}, respectively) (Table 2). CEC is high to very high (mean values from 15 to 30 cmol$^+$kg^{-1}, Table 2), and Mg is the dominant cation on the exchange complex.

Table 2. Comparison of soil variables between MINE and FIRE sites. Second and third columns present average values (±SD) for each variable and type of disturbed site. The last column indicates the *p*-values for each comparison. *P*-values higher than 0.05 are considered non-significant.

Soil Variable	Type of Disturbed Site		*p*-Value
	MINE	**FIRE**	
pH H_2O	7.89 (±0.59)	6.64 (±0.52)	<0.001
Soil water retention (g H_2O 100 g^{-1} soil)			
Wfc	26.2 (±9.4)	46.6 (±13.4)	0.008
Wwp	12.0 (±5.8)	32.6 (±12.8)	0.002
AWS	14.2 (±4.5)	14.0 (±5.5)	0.832
C and N (mass %)			
Total C	1.13 (±1.82)	6.62 (±3.09)	<0.001
Total N	0.05 (±0.04)	0.36 (±0.15)	<0.001
C/N ratio	19.1 (±13.8)	18.1 (±3.8)	0.573
Pseudo-total concentrations of major and trace elements (mg kg^{-1})			
P	81.1 (±65.0)	176 (±54)	0.005
S	283 (±479)	263 (±145)	0.121
Co	125 (±62.2)	434 (±172)	<0.001
Cr	1275 (±695)	5826 (±2202)	<0.001
Mn	1516 (±610)	4421 (±1082)	<0.001
Ni	1893 (±679)	2941 (±1082)	0.036
DTPA-extractable Ni (mg kg^{-1})	18.9 (±17.4)	155 (±62)	<0.001
P-Olsen (mg kg^{-1})	0.59 (±0.47)	2.83 (±1.85)	<0.001
CEC and exchangeable cations (cmol+ kg^{-1})			
CEC	15.9 (±10.1)	30.2 (±12.8)	0.024
Ca^{2+}	2.6 (±2.0)	9.0 (±5.5)	0.02
Mg^{2+}	10.3 (±6.8)	13.0 (±5.6)	0.319
K^+	0.1 (±0.1)	0.3 (±0.2)	0.004
Ca:Mg	0.4 (±0.3)	0.9 (±0.8)	0.052
Soil microbial activities (µg product $g^{-1}h^{-1}$)			
Urease	2.2 (±2.1)	5.4 (±2.7)	0.002
Arylsulphatase	5.3 (±7.0)	80.2 (±30.7)	<0.001
β-glucosidase	48.7 (±13.7)	82.6 (±20.4)	0.001
Alkaline phosphatase	17 (±16)	252 (±177)	<0.001
FDA hydrolysis	2.5 (±2.7)	40.0 (±14.5)	<0.001

PCA analysis of soil data identified four principal components (PC) which explained 86% of variance in soil properties. The first PC explained most of the variance (58%). It was negatively correlated to soil pH and positively correlated to Ni-DTPA and most fertility factors: water retention parameters (Wfc and Wwp), FDA hydrolysis and enzyme activities, CEC and exchangeable K and Ca, Olsen-P, and total concentrations of C and N (Figure 2a). Pseudototal concentrations of Co, Cr, and Mn were positively correlated to PC1 and PC2 (11% of total variance). Exchangeable Mg had a little contribution to both PCs, whereas pseudototal S concentration was not correlated to any of these PCs (Figure 2a).

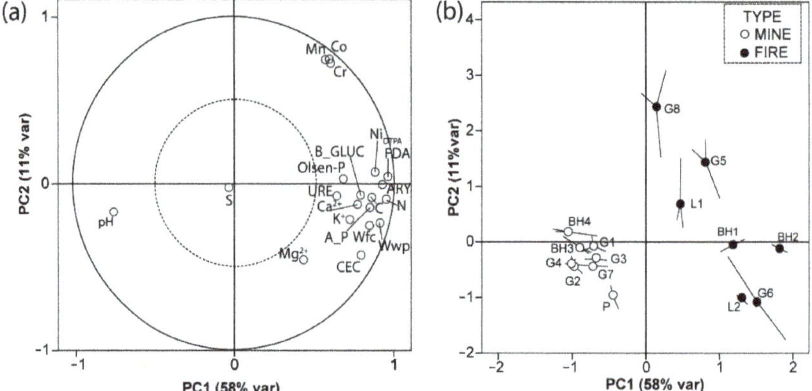

Figure 2. Principal component analysis (PCA) analysis of soil data. (**a**) Loadings of soil variables on principal components (PCs). Pseudototal concentrations of C, Co, Cr, Mn, N, and S are represented by their chemical symbol; A_P, Alkaline Phosphatase; ARY, Arylsulphatase; B_GLUC, β-glucosidase; CEC, cation exchange capacity; Ca^{2+}, K^+, Mg^{2+}, exchangeable concentrations of Ca, K, and Mg; FDA, hydrolysis of Fluorescein Diacetate; Ni_{DPTA}, available Ni; URE, Urease; Wfc, water at field capacity; Wwp: water at wilting point. (**b**) Scatterplot of soil samples (one point is the centroid of three samples per site, lines connect centroids with the position of each sample) on the two first PCs. Empty dots, MINE sites; black dots, FIRE sites. Codes indicate site: BH, Bukit Hampuan FR; G, Garas; L, Lompoyou; P, Paliu.

Projection of soil samples on first two PCs show a clear separation between FIRE and MINE plots. MINE plots are placed in a dense swarm on negative values for PC1 and ranging from −1 to 0.5 in PC2. FIRE samples had positive values in PC1 and were dispersed along PC2, from −1 to 3 (Figure 2b). Therefore, MINE plots had soils with higher pH and lower fertility than FIRE plots. However, the concentrations of trace elements were higher on FIRE than on MINE sites. Nested ANOVAs confirmed the inferences from PCA. For most of the analyzed variables, there were significant differences between site types (Table 2). It is remarkable that FIRE soils had better water retention properties (i.e., higher Wfc and Wwp) than MINE soils, but the available water storage (AWS) was similar between site types (around 14 g H_2O per 100 g of soil, Table 2).

3.2. Plant Communities

A total of 42 plant species were sampled in the 15 studied plots (Table A1). Plant cover in MINE sites was lower than in FIRE sites (45% vs. 99%, *p*-value < 0.001). Number of sampled species per plot ranged from 2 to 11, whereas Shannon's Index ranged from 0.11 to 1.93. These values were similar in MINE and in FIRE plots (Table 3). The vegetation in MINE plots was dominated by different grass species (*Paspalum* spp. and others) with incidental presence of pioneer trees such as *Neonauclea gigantea* or *Ceuthostoma terminale* (present only in MINE plots from Bukit Hampuan) (Table A1). In FIRE plots the fern *Pteridium esculentum* was dominant, with minor presence of grasses (*Imperata cylindrica*, *Miscanthus floridulus*), pioneer trees (*Trema* sp., *Vitex* spp.), and gingers (family Zingiberaceae, present only in FIRE plots from Bukit Hampuan) (Table A1). The Ni-hyperaccumulator (*Phyllanthus rufuschaneyi*) was found growing on FIRE plots only. Several alien species were also found: *Mimosa pudica* (only in MINE plots), *Lantana camara* (only in FIRE plots), or *Chromolaena odorata* (in MINE and in FIRE plots). Results of taxonomic NMDS ordination showed a clear separation between the communities in MINE and in FIRE plots (Figure 3a). This separation along NMDS1 is mainly correlated to variation in soil properties (summarized by PC1, the first principal component of soil PCA, Figure 3a), whereas the vectors altitude, aspect, and time since disturbance were correlated (along NMDS2) to differences between Bukit Hampuan (BH) and the other plots within types of disturbed sites.

Table 3. Summary of plant cover and taxonomic diversity (number of species and Shannon's H index) in sampled plots in areas disturbed by fire (FIRE sites) or by quarrying/soil excavation (MINE sites). For each variable, the mean and the minimum-maximum (within brackets), are presented. *p*-values of one-way ANOVAs are presented in the last column.

Variable	MINE Sites	FIRE Sites	*p*-Value
Plant cover (%)	45.1 (21–84)	99.5 (98–100)	<0.001
N of Species	5 (3–9)	6 (2–11)	0.322
Shannon's H	0.92 (0.18–1.93)	1.05 (0.11–1.63)	0.672

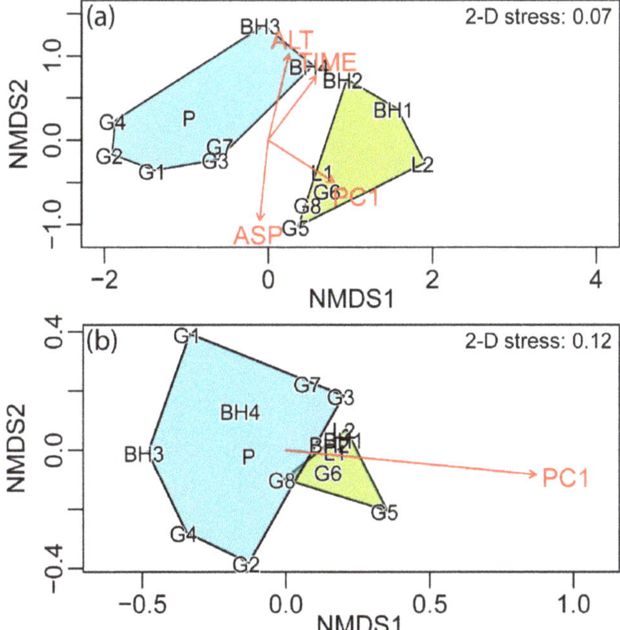

Figure 3. Non-metric multidimensional scaling (NMDS) of plant communities growing on degraded ultramafic areas in Sabah. (**a**) NMDS based on species composition. (**b**) NMDS based on community weighed means (CWMs) for 15 functional traits. Colored areas are convex hulls grouping MINE (blue polygons) and FIRE (green polygons) plots. The red arrows indicate environmental variables which are significantly correlated (i.e., *p*-value < 0.05) to the ordination. Abbreviations for environmental vectors are ALT, altitude above sea level; ASP, aspect; PC1, first principal component of soil PCA (see Figure 2); TIME, years from last disturbance. 2-D stress for each ordination is indicated in the graphs.

Regarding functional traits, similar values in CWM were obtained in both types of altered sites in 11 out of the 15 studied traits (Table 4). No variation was observed in life cycle (all sampled species were perennials). Depth of underground system was similar in the two types (0–30 cm). Plant height was less than 1 m in MINE plots as compared to 1–3 m in FIRE sites. Density of stems was higher in MINE sites, where communities were dominated by grasses/sedges. Plants with fixation of atmospheric N_2 were only present in MINE communities, whereas Ni hyperaccumulation was only present in FIRE plots, although low frequency of this trait made the differences between site types non-significant. Leaf traits (except Mn concentration) were similar between communities, whereas foliar Mn in MINE sites was two-fold higher than FIRE sites (Table 4). Compared to average values from the TRY database, our studied plant communities had lower SLA and lower foliar concentrations of N, P, Ca, and Mn, whereas foliar K were in similar ranges and foliar Mg was much higher.

Table 4. CWM for 15 functional traits assessed in our study. For information on the type of variable, units and categories of each trait, please refer to Table 1. The second and the third columns present the CWM for each trait averaged for each type of site (MINE vs. FIRE), followed by the standard deviation (in the case of binary or quantitative traits) or by the maximum and minimum (in the case of ordinal traits). The fourth column indicate the p-values. We considered significant differences when $p < 0.05$. For comparative purposes, the last two columns present mean values of certain traits from the TRY database [31] and from an extensive study of ultramafic flora of Sabah [32].

Trait	Type of Disturbed Site		p-Value	Mean in TRY Database [1]	Mean in Sabah Ultramafic Flora [2]
	MINE	FIRE			
Life cycle (binary)	1.0 (±0.0)	1.0 (±0.0)	1.0	-	-
Lateral spreading capacity (binary)	0.6 (±0.4)	0.7 (±0.2)	0.355	-	-
Depth root system (ordinal)	1.4 (1–2)	1.9 (1–2)	0.066	-	-
Plant height (ordinal)	3.4 (1–5)	5.0 (5–5)	0.015	-	-
Density of stems (ordinal)	2.1 (1–3)	1.1 (1–2)	0.006	-	-
N_2 fixation (binary)	0.1 (±0.1)	0.0 (±0.0)	0.016	-	-
Ni_Hyperaccum (binary)	0.0 (±0.0)	0.01 (±0.02)	0.285	-	-
SLA ($mm^2\ mg^{-1}$)	11.2 (±6.0)	8.3 (±3.0)	0.270	16.6	-
Leaf N ($mg\ g^{-1}$)	10.3 (±4.7)	11.7 (±2.6)	0.500	17.4	-
Leaf P ($mg\ g^{-1}$)	0.78 (±0.38)	0.88 (±0.26)	0.592	1.23	0.41
Leaf K ($mg\ g^{-1}$)	8.6 (±4.7)	11.6 (±1.3)	0.135	8.4	3.8
Leaf Ca ($mg\ g^{-1}$)	3.23 (±1.35)	2.63 (±1.01)	0.354	9.05	6.36
Leaf Mg ($mg\ g^{-1}$)	4.07 (±2.24)	3.13 (±1.77)	0.387	2.61	3.03
Leaf Mn ($\mu g\ g^{-1}$)	68.4 (±49.4)	30.6 (±18.9)	0.047	189	588
Leaf Ca:Mg	0.60 (±0.30)	0.57 (±0.20)	0.827	-	-

[1] TRY is an international database on plant functional traits. In 2011, it contained almost 3 million trait data entries for around 69,000 plant species [31]; [2] average values of flora from several undisturbed ultramafic sites of Sabah.

NMDS ordination of the studied plots on the basis of CWM showed that the FIRE and MINE plots partially overlap (Figure 3b). FIRE plots occupy less area (i.e., CWMs were similar) in the NMDS space, probably due to the dominance of *Pteridium esculentum* in those communities. Interestingly, only the soil conditions (summarized as the soil principal component PC1) are significantly correlated to this ordination. This fact indicates that the small differences in CWMs between communities are related to differences in soil properties.

4. Discussion

4.1. Soil Properties in Disturbed Ultramafic Habitats

In our study, we focused on tropical ultramafic areas disturbed by fire and excavation. Soil formation from serpentinite bedrock under tropical conditions leads to cambisols or to cambic leptosols with neutral to basic pH, very high CEC, high total and exchangeable Mg concentration, and high Ni availability [38,51]. In those soils, nutrients are scarce and mostly distributed in the upper soil horizons, where they are kept by an intense recycling of decaying organic matter [51]. The soils in MINE and FIRE sites of our study have the typical ultramafic properties described in previous lines. However, intense disturbance in MINE plots has eliminated the richer topsoil and resulted in a strong reduction of chemical fertility. Thus, soil carbon concentration is five-fold lower in MINE than in FIRE plots, the CEC of MINE plots is half of FIRE plots, the concentration of important nutrients is two-(P, K) to seven-(N) fold lower in MINE than in FIRE soils and the soil biological activity is extremely low in MINE plots. Soils derived from serpentinites are rich in 2:1 clays (smecites) [51] which have good water retention capacities. These clays could be responsible for the similar available water storage in MINE and FIRE soils, despite the important differences in soil organic matter content. Different studies have shown that ultramafic plants play a major role in the building up of Ni concentrations in the topsoil and its maintenance through biogeochemical recycling [51,52]. This phenomenon of plant recycling could be responsible for the high pseudototal concentrations of trace elements (Ni, Cr, Co, Mn) and phytoavailable Ni in FIRE soils.

4.2. Functional Traits in Disturbed Ultramafic Habitats

Plants must deal with extreme conditions in ultramafic soils, which are even harsher in human-disturbed soils [3,4,11]. Our results on CWM for different functional traits allow the identification of main plant community adaptations to ultramafic stress. Despite the differences in soil conditions and taxonomic composition between MINE and FIRE sites, the CWM for most of the studied traits were similar in both types of disturbed areas. Moreover, the small differences we found were correlated to changes in soil properties (see 'functional NMDS'). In both types of disturbances, the plant communities had a conservative strategy (e.g., slow-growing species that conserve resources) [53]. As an example, CWM values of SLA are below the average SLA value in TRY database [31], and in the lower range of SLA values reported for tree species from tropical forest in Mount Kinabalu (SLA from 2.72 to 120.3 mm^2 mg^{-1}) [54] and for herbaceous plants growing on ultramafic substrates from Lesbos island (East Mediterranean) (SLA from around 10 to 45 mm^2 mg^{-1}) [55]. SLA has been correlated with relative growth rate and stress tolerance, as well as a protection against herbivores [56,57], thus we can conclude that our studied communities are characterized by low relative growth rates but high tolerance to stress. Community mean leaf concentrations of N, P, Ca, and Mn were slightly below the average values from TRY database and (for N and P) below the vegetation from Lesbos island [31,55]. In contrast, CWM values for leaf K and Mg are over the mean for TRY. High Mg CWM are understandable because Mg is the dominant element in exchange complex in serpentine-derived soils [38,51] and in our samples. However, leaf K concentrations (average 10.1 mg g^{-1} K) are unusually high if we consider that the soils in the studied plots were deficient in this element. It is known that serpentinite-derived soils from Sabah are well drained and prone to drought [36,58]. Potassium plays a role in plant tolerance to drought [59], so these high levels of potassium could be an adaptation of pioneer ultramafic flora to cope with water limitation.

Regarding N and P, plant communities had intermediate leaf concentrations (11 and 0.83 mg g^{-1} of N and P, respectively), which are lower than those observed in temperate ultramafic plants from Lesbos [55]. However, these concentrations are remarkable considering: (i) the extremely low soil concentrations of these nutrients, especially P (average N soil concentration 2.05 mg g^{-1}; average P-Olsen 1.7 mg kg^{-1}); (ii) the lower average leaf P concentration (0.41 mg g^{-1}) in plants from undisturbed ultramafic areas of Sabah [32]. Therefore, we may conclude that the plant communities in our studied plots are characterized by a high capacity of nutrient absorption and storage. Our findings in tropical degraded ultramafic areas are congruent with some of the functional characteristics previously described in ultramafic flora in temperate regions: slow growth rates, high investment in anti-herbivore defense, storage of nutrients, and efficient nutrient use [4,55]. This nutrient-conservative strategy is not restricted to ultramafic flora, but it is usually found in nutrient-limited ecosystems [53].

Soil Ca:Mg molar ratio has been identified as one of the important factors involved in the infertility of ultramafic soils, due to the antagonistic effect of high Mg concentrations over Ca uptake by plants [3,4]. In fact, the ability to maintain a leaf Ca:Mg molar ratio > 1 has been indicated as an important trait to explain adaptation to ultramafic soils in different ultramafic plants from temperate regions [60,61]. However, increased Mg requirements have been found in several temperate ultramafic plant ecotypes [3]. In contrast, leaf Ca:Mg molar ratios in our studied communities were around 0.5 (i.e., almost double of Mg moles in comparison with Ca moles in the leaves). Our results may be explained either by efficient tolerance mechanisms to excess Mg in plant tissues or higher Mg requirements in tropical ultramafic plants. However, more research is needed to clarify this topic.

The plant communities in MINE and FIRE plots differed in only four out of 15 studied traits: plant height, density of stems, N$_2$ fixation, and leaf Mn concentration. Plant height was higher in FIRE plots as a result of soils with more nutrient resources, but also as a response to high competition for light in sites with dense plant cover. The higher density of stems in MINE plots is just a consequence of the dominance of grasses and sedges—with high number of culms—in these sites. N_2 fixation is an important characteristic, especially in nutrient-poor soils. Thus, it was unexpected that this trait appeared in very low frequency, and only in MINE sites. The difference in leaf manganese concentration is interesting as it cannot be explained by differences in soil Mn concentration: MINE sites had lower soil Mn and higher leaf Mn concentration than FIRE sites. Leaf Mn concentration has been proposed as a proxy to phosphorus-acquisition efficiency [62]: mechanisms for phosphorus mobilization at root level (release of carboxylates) provoke a significant increase in the absorbed Mn. However, to our knowledge, only one species from MINE plots (*Ceuthostoma terminale*, Casuarinaceae, with clusteroid roots) clearly has this strategy. The dominant species in FIRE plots (the fern, *Pteridium esculentum*) has very low leaf Mn concentrations (values from 5.5 to 11.5 mg kg^{-1}). Therefore, a more detailed study on the mechanisms of P-absorption would be needed to determine if this difference in Mn concentration in disturbed ultramafic communities is related to enhanced carboxylate release or whether it is just an artifact caused by the dominance of *P. esculentum*.

4.3. Implication in Revegetation of Ultramafic Degraded Areas

The analysis of functional composition of plant communities on degraded metal-rich soils may provide clues for the restoration of those areas, as well as trait-assisted selection of potential species for revegetation.

In our study we analyzed two types of tropical anthropized ultramafic habitats: serpentinite quarries or dumpsites and burnt areas. Soil conditions in these habitats mimic two possible ultramafic post-mining scenarios: (*a*) raw serpentine bedrock or saprolite exposed after soil excavation and (*b*) serpentine tailing amended with organic matter or covered with topsoil. On the basis of similar functional composition of tropical plant communities in FIRE and MINE habitats, we can conclude that suitable species for revegetation of scenarios *a* and *b* should possess the following attributes: perennial life cycle, lateral spreading capacity, rooting depth lower than 30 cm, and a nutrient conservative strategy (low SLA, high leaf K, and intermediate N and P concentrations). However, plants for scenario *a* should have between 10 and 30 stems per dm^2 and height between 0.3 and 0.59 m. In contrast, for scenario *b* the number of stems should be between 1 and 10 per dm^2 and the preferred height between 1 and 3 m. Finally, due to extremely low nutrient concentrations, revegetation in scenario *a* must include species with the ability of N_2 fixation [63,64].

Although some of these traits are present also in ultramafic plants from temperate areas, our results can be extrapolated only to tropical areas due to differences in serpentinite soils from tropical and temperate regions [51] and the specific conditions of tropical climate (e.g., high pluviometry, stable temperature, lack of marked dry season). As an example, high biomass producing temperate Ni-hyperaccumulators of the genus *Odontarrhena* spp. (formerly *Alyssum*) cultivated in a phytomining trial in Sulawesi (Indonesia) showed a low Ni concentration and a reduced biomass production [14].

Our results indicate that in case of extremely anthropized soil—scenario *a*—cultivation of selected plants in nursery and planting would be necessary to obtain a good plant cover (average plant cover in MINE plots is around 48%), whereas in scenario *b* spontaneous plant colonization may be an effective strategy for revegetation (almost 100% of plant cover in FIRE plots). However, spontaneous revegetation is feasible only if plant populations are locally present to act as seed sources [63] and if alien species are absent [65].

The use of species that fulfil the previous criteria in restoration of ultramafic degraded areas will reduce erosion (perennial plant cover, increased covered area due to lateral spreading capacity, limitation of surface runoff and sediment trapping due to dense number of stems) and nutrients will be conserved in the site. However, the obtained plant community will be poor in terms of functional diversity. Different studies have shown that the use of species with complementary functional traits have positive synergistic effects on the phytoremediation of multicontaminated soils and the reclamation of extremely degraded mine tailings [28,29]. With this aim, three additional elements should be explored to ameliorate restoration approaches in ultramafic habitats.

Nutrient cycling: plants with conservative strategies tend to have leaves with low SLA and to recover nutrients (i.e., nutrient resorption) during leaf senescence. Thus, produced litter is difficult to degrade and it contains reduced nutrient concentrations [23]. Measures to increase nutrient recycling, such as the introduction of N_2-fixing plants or inoculation of soil fauna (i.e., earthworms) should be explored [63,66].

Managing competition: the use of perennial plants with lateral spreading capacity can imply in some cases the selection of strong competitive species. Use of this species can create a favorable habitat for the colonization of other species (i.e., they can act as nurse species) or can lead to the creation of a dense monospecific plant cover that outcompetes any other species (e.g., degraded grasslands of *Imperata cylindrica* [34]). Depending on the desired outcomes, management would be needed to reduce competitive pressure of the dominant species. For instance, a sacrifice fallow crop with the exotic fast-growing *Acacia mangium* showed to be useful suppressing dominant *Imperata cylindrica* and creating microconditions for the germination and growth of native tree species [67].

Ni-hyperaccumulation: Ni-phytomining using hyperaccumulator plants has been proposed as being compatible with the restoration of mined areas [68]. Nickel would be recovered from the biomass of cultivated native hyperaccumulators and the obtained incomes would be used to cover (part of) restoration costs. The absence of Ni-hyperaccumulation from MINE plots indicates that phytomining would only be feasible in scenario *b*. A further improvement would consist of the selection of hyperaccumulators with other traits that fit conditions in degraded ultramafic areas (e.g., resprouting ability, resistance to drought, and full sunlight).

Author Contributions: Study design, C.Q.-S., M.-P.F., G.E. and S.L.; fieldwork organization and sampling, C.Q.-S., R.R., J.B.S. and R.N.; laboratory and formal analysis, C.Q.-S.; writing—original draft preparation, C.Q.-S.; writing—review and editing, all authors; supervision, C.Q.-S., G.E. and S.L.; funding acquisition, G.E. and S.L. All authors have read and agreed to the published version of the manuscript.

Funding: C.Q-S. contract and this research were funded by the French National Research Agency (ANR) through the national program "Investissements d'avenir" with the reference ANR-10-LABX-21-01/LABEX RESSOURCES21.

Acknowledgments: Sabah Biodiversity Centre provided C.Q-S. with the Access License ref no: JKM/MBS.1000-2/2 JLD.4(101). Sabah Parks and Sabah Forestry Department provided specific permits for the access to sampling sites. Romain Goudon, Lucas Charrois, and Stephane Colin are acknowledged for their contribution to the analyses of soils. Two anonymous reviewers provided valuable comments for the improvement of the manuscript.

Conflicts of Interest: The authors declare no conflict of interest. The funders had no role in the design of the study; in the collection, analyses, or interpretation of data; in the writing of the manuscript, or in the decision to publish the results.

Appendix A

Table A1. List of plant species identified in 15 plots in degraded ultramafic areas from Sabah (Malaysia). Species codes are based in the first two letters of the genera and the specific epithet. When species identification was not possible, the codes correspond to the growing form: *F*, forb; *FE*, fern; *G*, grass; *T*, tree; *ZI*, ginger. Plant division is indicated in the third column when the plant family is not known. The last two columns indicate the number of plots where each species is present, and the relative cover (%) of that species averaged by the number of plots where it is present.

CODE	Species	Division/Family	N Occurrences/Average Cover (%)	
			FIRE plots	MINE plots
CETE	*Ceuthostoma terminale*	Casuarinaceae	0/0	2/4
CHOD	*Chromolaena odorata*	Asteraceae	4/26	2/3
CLSP	*Clausena* sp.	Rutaceae	1/10	0/0

Table A1. *Cont.*

CODE	Species	Division/Family	N Occurrences/Average Cover (%)	
			FIRE plots	MINE plots
COSP	*Colona* sp.	Malvaceae	1/7	0/0
COMSP	*Commersonia* sp.	Malvaceae	1/1	1/3
CY#01	*Cyperus* sp.	Cyperaceae	1/4	0/0
DEFR	*Decaspermom fruticosum*	Myrtaceae	1/1	0/0
ETCO	*Etlingera coccinea*	Zingiberaceae	1/4	0/0
F#01	-	Dicotyledon	0/0	1/3
FE#01	-	Polypodiophyta	0/0	2/2
FE#02	-	Polypodiophyta	0/0	1/8
FISP	*Fimbristylis* sp.	Cyperaceae	0/0	4/35
G#01	-	Poaceae	0/0	1/24
G#02	-	Poaceae	0/0	3/24
G#03	-	Poaceae	0/0	1/1
G#04	-	Poaceae	0/0	1/26
G#05	-	Poaceae	0/0	1/3
G#06	-	Poaceae	0/0	1/3
IMCY	*Imperata cylindrica*	Poaceae	4/5	1/8
LACA	*Lantana camara*	Verbenaceae	1/10	0/0
LYSP	*Lygodium* sp.	Lygodiacaeae	3/4	0/0
MA#01	*Macaranga* sp.1	Euphorbiaceae	1/1	0/0
MA#02	*Macaranga* sp.2	Euphorbiaceae	1/4	0/0
MA#03	*Macaranga* sp.3	Euphorbiaceae	1/2	0/0
ME#01	*Melastoma* sp.	Melastomataceae	1/7	0/0
ME#02	*Medinilla* sp.	Melastomataceae	0/0	1/1
MIFL	*Miscanthus floridulus*	Poaceae	3/9	2/5
MIPU	*Mimosa pudica*	Fabaceae	0/0	3/5
NASP	*Nauclea* sp.	Rubiaceae	1/11	0/0
NEGI	*Neonauclea gigantea*	Rubiaceae	1/2	2/2
PASP1	*Paspalum* sp1.	Poaceae	0/0	2/4
PASP2	*Paspalum* sp2.	Poaceae	0/0	6/3
PHRU	*Phyllanthus rufuschaneyi*	Phyllanthaceae	1/5	0/0
PTES	*Pteridium esculentum*	Dennstaedtiaceae	7/63	0/0
RU#01	*Rubus* sp.	Rosaceae	0/0	1/1
T#01	-	Dicotyledon	1/2	0/0
T#02	-	Dicotyledon	1/1	0/0
T#03	-	Dicotyledon	1/3	0/0
TRSP	*Trema* sp.	Cannabaceae	2/3	0/0
VIPI	*Vitex pinnata*	Lamiaceae	1/5	0/0
VISP	*Vitex* sp.	Lamiaceae	1/5	0/0
ZI#01	-	Zingiberaceae	1/4	0/0

References

1. Kruckeberg, A.R. An essay: Geoedaphics and island biogeography for vascular plants. *Aliso: A J. Syst. Evolut. Bot.* **1991**, *13*, 225–238. [CrossRef]
2. van der Ent, A.; Erskine, P.D.; Mulligan, D.R.; Repin, R.; Karim, R. Vegetation on ultramafic edaphic "islands" in Kinabalu Park (Sabah, Malaysia) in relation to soil chemistry and elevation. *Plant Soil* **2016**, *403*, 77–101. [CrossRef]
3. Brady, K.U.; Kruckeberg, A.R.; Bradshaw, H.D., Jr. Evolutionary ecology of plant adaptation to serpentine soils. *Annu. Rev. Ecol. Evolut. Syst.* **2005**, *36*, 243–266. [CrossRef]
4. Kazakou, E.; Dimitrakopoulos, P.G.; Baker, A.J.M.; Reeves, R.D.; Troumbis, A.Y. Hypotheses, mechanisms and trade-offs of tolerance and adaptation to serpentine soils: From species to ecosystem level. *Biol. Rev.* **2008**, *83*, 495–508. [CrossRef]
5. Reeves, R.D.; Baker, A.J.M.; Borhidi, A.; Berazain, R. Nickel Hyperaccumulation in the Serpentine Flora of Cuba. *Ann. Bot.* **1999**, *83*, 29–38. [CrossRef]
6. Isnard, S.; L'huillier, L.; Rigault, F.; Jaffré, T. How did the ultramafic soils shape the flora of the New Caledonian hotspot? *Plant soil* **2016**, *403*, 53–76. [CrossRef]
7. Galey, M.L.; Van der Ent, A.; Iqbal, M.C.M.; Rajakaruna, N. Ultramafic geoecology of South and Southeast Asia. *Botanical Stud.* **2017**, *58*, 18. [CrossRef]
8. van der Ent, A.; Repin, R.; Sugau, J.; Wong, K.M. Plant diversity and ecology of ultramafic outcrops in Sabah (Malaysia). *Aust. J. Bot.* **2015**, *63*, 204–215. [CrossRef]
9. Harrison, S.; Rajakaruna, N. *Serpentine: The evolution and ecology of a model system*; Univ of California Press: Berkeley, CA, USA, 2011; ISBN 0-520-26835-0.
10. Whiting, S.N.; Reeves, R.D.; Richards, D.; Johnson, M.S.; Cooke, J.A.; Malaisse, F.; Paton, A.; Smith, J.A.C.; Angle, J.S.; Chaney, R.L. Research priorities for conservation of metallophyte biodiversity and their potential for restoration and site remediation. *Restor. Ecol.* **2004**, *12*, 106–116. [CrossRef]
11. O'Dell, R.E.; Claassen, V.P. Serpentine revegetation: A review. *Northeast. Nat.* **2009**, *16*, 253–271. [CrossRef]
12. Mudd, G.M. Global trends and environmental issues in nickel mining: Sulfides versus laterites. *Ore Geol. Rev.* **2010**, *38*, 9–26. [CrossRef]
13. Losfeld, G.; L'Huillier, L.; Fogliani, B.; Jaffré, T.; Grison, C. Mining in New Caledonia: Environmental stakes and restoration opportunities. *Environ. Sci. Pollut. Res.* **2015**, *22*, 5592–5607. [CrossRef] [PubMed]
14. van der Ent, A.; Baker, A.J.M.; Van Balgooy, M.M.J.; Tjoa, A. Ultramafic nickel laterites in Indonesia (Sulawesi, Halmahera): Mining, nickel hyperaccumulators and opportunities for phytomining. *J. Geochem. Explor.* **2013**, *128*, 72–79. [CrossRef]
15. Sodhi, N.S.; Koh, L.P.; Brook, B.W.; Ng, P.K. Southeast Asian biodiversity: An impending disaster. *Trends Ecol. Evol.* **2004**, *19*, 654–660. [CrossRef] [PubMed]
16. van der Ent, A. The ecology of ultramafic areas in Sabah: Threats and conservation needs. *Gardens' Bull. Singap.* **2011**, *63*, 385–393.
17. Pfeifer, M.; Kor, L.; Nilus, R.; Turner, E.; Cusack, J.; Lysenko, I.; Khoo, M.; Chey, V.K.; Chung, A.C.; Ewers, R.M. Mapping the structure of Borneo's tropical forests across a degradation gradient. *Remote Sens. Environ.* **2016**, *176*, 84–97. [CrossRef]
18. McCoy, S.; Jaffré, T.; Rigault, F.; Ash, J.E. Fire and succession in the ultramafic maquis of New Caledonia. *J. Biogeogr.* **1999**, *26*, 579–594. [CrossRef]
19. Garnier, É.; Navas, M.-L. *Diversité Fonctionnelle des Plantes: Traits des Organismes, Structure des Communautés, Propriétés des écosystèmes: Cours*; De Boeck: Bruxelles, Belgium, 2013; ISBN 2-8041-7562-6.
20. Fukami, T.; Bezemer, B.T.; Mortimer, S.R.; van der Putten, W.H. Species divergence and trait convergence in experimental plant community assembly. *Ecol. Lett.* **2005**, *8*, 1283–1290. [CrossRef]
21. Garnier, E.; Cortez, J.; Billès, G.; Navas, M.-L.; Roumet, C.; Debussche, M.; Laurent, G.; Blanchard, A.; Aubry, D.; Bellmann, A. Plant functional markers capture ecosystem properties during secondary succession. *Ecology* **2004**, *85*, 2630–2637. [CrossRef]
22. Lange, B.; Faucon, M.-P.; Delhaye, G.; Hamiti, N.; Meerts, P. Functional traits of a facultative metallophyte from tropical Africa: Population variation and plasticity in response to cobalt. *Environ. al Exp. Bot.* **2017**, *136*, 1–8. [CrossRef]

23. De Deyn, G.B.; Cornelissen, J.H.; Bardgett, R.D. Plant functional traits and soil carbon sequestration in contrasting biomes. *Ecol. Lett.* **2008**, *11*, 516–531. [CrossRef] [PubMed]

24. Faucon, M.-P.; Houben, D.; Lambers, H. Plant functional traits: Soil and ecosystem services. *Trends Plant Sci.* **2017**, *22*, 385–394. [CrossRef] [PubMed]

25. Ilunga wa Ilunga, E.; Mahy, G.; Piqueray, J.; Seleck, M.; Shutcha, M.N.; Meerts, P.; Faucon, M.-P. Plant functional traits as a promising tool for the ecological restoration of degraded tropical metal-rich habitats and revegetation of metal-rich bare soils: A case study in copper vegetation of Katanga, DRC. *Ecol. Eng.* **2015**, *82*, 214–221. [CrossRef]

26. Gilardelli, F.; Sgorbati, S.; Armiraglio, S.; Citterio, S.; Gentili, R. Ecological filtering and plant traits variation across quarry geomorphological surfaces: Implication for restoration. *Environ. al Manag.* **2015**, *55*, 1147–1159. [CrossRef]

27. Horáčková, M.; Řehounková, K.; Prach, K. Are seed and dispersal characteristics of plants capable of predicting colonization of post-mining sites? *Environ. al Sci. Pollut. Res.* **2016**, *23*, 13617–13625. [CrossRef]

28. Desjardins, D.; Brereton, N.J.; Marchand, L.; Brisson, J.; Pitre, F.E.; Labrecque, M. Complementarity of three distinctive phytoremediation crops for multiple-trace element contaminated soil. *Sci. Total Environ.* **2017**, *610*, 1428. [CrossRef]

29. Navarro-Cano, J.A.; Goberna, M.; Verdú, M. Using plant functional distances to select species for restoration of mining sites. *J. Appl. Ecol.* **2019**, *56*, 2353–2362. [CrossRef]

30. Lohbeck, M.; Poorter, L.; Lebrija-Trejos, E.; Martínez-Ramos, M.; Meave, J.A.; Paz, H.; Pérez-García, E.A.; Romero-Pérez, I.E.; Tauro, A.; Bongers, F. Successional changes in functional composition contrast for dry and wet tropical forest. *Ecology* **2013**, *94*, 1211–1216. [CrossRef]

31. Kattge, J.; Diaz, S.; Lavorel, S.; Prentice, I.C.; Leadley, P.; Bönisch, G.; Garnier, E.; Westoby, M.; Reich, P.B.; Wright, I.J. TRY–a global database of plant traits. *Glob. Chang. Biol.* **2011**, *17*, 2905–2935. [CrossRef]

32. van der Ent, A.; Mulligan, D.R.; Repin, R.; Erskine, P.D. Foliar elemental profiles in the ultramafic flora of Kinabalu Park (Sabah, Malaysia). *Ecol. Res.* **2018**, *33*, 659–674. [CrossRef]

33. Proctor, J.; Lee, Y.F.; Langley, A.M.; Munro, W.R.C.; Nelson, T. Ecological studies on Gunung Silam, a small ultrabasic mountain in Sabah, Malaysia. I. Environment, forest structure and floristics. *J. Ecol.* **1988**, 320–340. [CrossRef]

34. Woods, P. Effects of logging, drought, and fire on structure and composition of tropical forests in Sabah, Malaysia. *Biotropica* **1989**, *21*, 290–298. [CrossRef]

35. Kitayama, K.; Aiba, S.-I.; Majalap-Lee, N.; Ohsawa, M. Soil nitrogen mineralization rates of rainforests in a matrix of elevations and geological substrates on Mount Kinabalu, Borneo. *Ecol. Res.* **1998**, *13*, 301–312. [CrossRef]

36. Aiba, S.; Kitayama, K. Structure, composition and species diversity in an altitude-substrate matrix of rain forest tree communities on Mount Kinabalu, Borneo. *Plant Ecol.* **1999**, *140*, 139–157. [CrossRef]

37. IUSS Working Group WRB. *World Reference Base for Soil Resources 2014, update 2015 International soil classification system for naming soils and creating legends for soil maps. World Soil Resources Reports No 106*; World Soil Resources Reports; Food and Agriculture Organization of the United Nations: Rome, Italy, 2015.

38. van der Ent, A.; Cardace, D.; Tibbett, M.; Echevarria, G. Ecological implications of pedogenesis and geochemistry of ultramafic soils in Kinabalu Park (*Malaysia*). *Catena* **2018**, *160*, 154–169. [CrossRef]

39. Bandick, A.K.; Dick, R.P. Field management effects on soil enzyme activities. *Soil Biol. Biochem.* **1999**, *31*, 1471–1479. [CrossRef]

40. Adam, G.; Duncan, H. Development of a sensitive and rapid method for the measurement of total microbial activity using fluorescein diacetate (FDA) in a range of soils. *Soil Biol. Biochem.* **2001**, *33*, 943–951. [CrossRef]

41. Quintela-Sabarís, C.; Auber, E.; Sumail, S.; Masfaraud, J.-F.; Faucon, M.-P.; Watteau, F.; Saad, R.F.; van der Ent, A.; Repin, R.; Sugau, J.; et al. Recovery of ultramafic soil functions and plant communities along an age-gradient of the actinorhizal tree *Ceuthostoma terminale* (Casuarinaceae) in Sabah (Malaysia). *Plant Soil* **2019**, *440*, 201–218. [CrossRef]

42. Bruand, A.; Duval, O.; Gaillard, H.; Darthout, R.; Jamagne, M. Variabilité des propriétés de rétention en eau des sols: Importance de la densité apparente. *Etude et Gestion des sols* **1996**, *31*, 27–40.

43. Aran, D.; Maul, A.; Masfaraud, J.-F. A spectrophotometric measurement of soil cation exchange capacity based on cobaltihexamine chloride absorbance. *Comptes Rendus Geosci.* **2008**, *340*, 865–871. [CrossRef]

44. Olsen, S.R.; Cole, C.V.; Watanabe, F.S.; Dean, L.A. *Estimation of Available Phosphorus in Soils by Extraction with Sodium Bicarbonate*; USDA Circ.; US Gov. Print. Office: Washington, DC, USA, 1954.

45. Lindsay, W.L.; Norvell, W.A. Development of a DTPA soil test for zinc, iron, manganese, and copper. *Soil Sci. Soc. Am. J.* **1978**, *42*, 421–428. [CrossRef]

46. Schneider, C.A.; Rasband, W.S.; Eliceiri, K.W. NIH Image to ImageJ: 25 years of image analysis. *Nat. Methods* **2012**, *9*, 671. [CrossRef] [PubMed]

47. Pérez-Harguindeguy, N.; Díaz, S.; Garnier, E.; Lavorel, S.; Poorter, H.; Jaureguiberry, P.; Bret-Harte, M.S.; Cornwell, W.K.; Craine, J.M.; Gurvich, D.E. New handbook for standardised measurement of plant functional traits worldwide. *Aust. J. Bot.* **2013**, *61*, 167–234. [CrossRef]

48. Oksanen, J.; Blanchet, G.; Friendly, M.; Kindt, R.; Legendre, P.; McGlinn, D.; Minchin, P.R.; O'Hara, R.; Simpson, G.L.; Solymos, P.; et al. Vegan: Community Ecology Package; R Package Version 2.4-4. 2017. Available online: http://CRAN.R-project.org/package=vegan (accessed on 1 June 2020).

49. Lavorel, S.; Grigulis, K.; McIntyre, S.; Williams, N.S.; Garden, D.; Dorrough, J.; Berman, S.; Quétier, F.; Thébault, A.; Bonis, A. Assessing functional diversity in the field–methodology matters! *Funct. Ecol.* **2008**, *22*, 134–147. [CrossRef]

50. Laliberté, E.; Legendre, P. A distance-based framework for measuring functional diversity from multiple traits. *Ecology* **2010**, *91*, 299–305. [CrossRef]

51. Echevarria, G. Genesis and behaviour of ultramafic soils and consequences for nickel biogeochemistry. In *Agromining: Extracting Unconventional Resources from Plants*; Van der Ent, A., Echevarria, G., Baker, A.J.M., Morel, J.-L., Eds.; Mineral Resource Reviews Series; Springer Nature: Cham, Switzerland, 2018; pp. 135–156.

52. Estrade, N.; Cloquet, C.; Echevarria, G.; Sterckeman, T.; Deng, T.; Tang, Y.; Morel, J.-L. Weathering and vegetation controls on nickel isotope fractionation in surface ultramafic environments (Albania). *Earth Planet. Sci. Lett.* **2015**, *423*, 24–35. [CrossRef]

53. Reich, P.B. The world-wide 'fast–slow' plant economics spectrum: A traits manifesto. *J. Ecol.* **2014**, *102*, 275–301. [CrossRef]

54. Hidaka, A.; Kitayama, K. Allocation of foliar phosphorus fractions and leaf traits of tropical tree species in response to decreased soil phosphorus availability on Mount Kinabalu, Borneo. *J. Ecol.* **2011**, *99*, 849–857. [CrossRef]

55. Adamidis, G.C.; Kazakou, E.; Fyllas, N.M.; Dimitrakopoulos, P. Species adaptive strategies and leaf economic relationships across serpentine and non-serpentine habitats on Lesbos, eastern Mediterranean. *PLoS ONE* **2014**, *9*, e96034. [CrossRef]

56. Weiher, E.; van der Werf, A.; Thompson, K.; Roderick, M.; Garnier, E.; Eriksson, O. Challenging Theophrastus: A common core list of plant traits for functional ecology. *J. Veg. Sci.* **1999**, *10*, 609–620. [CrossRef]

57. Cornelissen, J.H.C.; Lavorel, S.; Garnier, E.; Diaz, S.; Buchmann, N.; Gurvich, D.E.; Reich, P.B.; ter Steege, H.; Morgan, H.D.; van der Heijden, M.G.A.; et al. A handbook of protocols for standardised and easy measurements of plant functional traits worldwide. *Aust. J. Bot.* **2003**, *51*, 335–380. [CrossRef]

58. van der Ent, A.; Wood, J.J. Orchids of extreme serpentinite (ultramafic) habitats in Kinabalu Park. *Males. Orchid J.* **2013**, *12*, 39–54.

59. Wang, M.; Zheng, Q.; Shen, Q.; Guo, S. The critical role of potassium in plant stress response. *International J. Mol. Sci.* **2013**, *14*, 7370–7390. [CrossRef] [PubMed]

60. O'Dell, R.E.; James, J.J.; Richards, J.H. Congeneric serpentine and nonserpentine shrubs differ more in leaf Ca: Mg than in tolerance of low N, low P, or heavy metals. *Plant Soil* **2006**, *280*, 49–64. [CrossRef]

61. Quintela-Sabarís, C.; Marchand, L.; Smith, J.A.C.; Kidd, P.S. Using AFLP genome scanning to explore serpentine adaptation and nickel hyperaccumulation in Alyssum serpyllifolium. *Plant and Soil* **2017**, *416*, 391–408. [CrossRef]

62. Lambers, H.; Hayes, P.E.; Laliberté, E.; Oliveira, R.S.; Turner, B.L. Leaf manganese accumulation and phosphorus-acquisition efficiency. *Trends Plant sci.* **2015**, *20*, 83–90. [CrossRef] [PubMed]

63. Bradshaw, A. The use of natural processes in reclamation—advantages and dificulties. *Landsc. Urban Plan.* **2000**, *51*, 89–100. [CrossRef]

64. Bradshaw, A. Restoration of mined lands—using natural processes. *Ecol. Eng.* **1997**, *8*, 255–269. [CrossRef]

65. Quintela-Sabarís, C.; Masfaraud, J.-F.; Séré, G.; Sumail, S.; van der Ent, A.; Repin, R.; Sugau, J.; Nilus, R.; Echevarria, G.; Leguédois, S. Effects of reclamation effort on the recovery of ecosystem functions of a tropical degraded serpentinite dump site. *J. Geochem. Explor.* **2019**, *200*, 139–151. [CrossRef]

66. Snyder, B.A.; Hendrix, P.F. Current and potential roles of soil macroinvertebrates (earthworms, millipedes and isopods) in ecological restoration. *Restor. Ecol.* **2008**, *16*, 629–636. [CrossRef]

67. Kuusipalo, J.; Ådjers, G.; Jafarsidik, Y.; Otsamo, A.; Tuomela, K.; Vuokko, R. Restoration of natural vegetation in degraded Imperata cylindrica grassland: Understorey development in forest plantations. *J. Veg. Sci.* **1995**, *6*, 205–210. [CrossRef]

68. Nkrumah, P.N.; Baker, A.J.; Chaney, R.L.; Erskine, P.D.; Echevarria, G.; Morel, J.L.; van der Ent, A. Current status and challenges in developing nickel phytomining: An agronomic perspective. *Plant Soil* **2016**, *406*, 55–69. [CrossRef]

Article

Effect of *Casuarina* Plantations Inoculated with Arbuscular Mycorrhizal Fungi and *Frankia* on the Diversity of Herbaceous Vegetation in Saline Environments in Senegal

Pape Ibrahima Djighaly [1,2,3,4,*], Daouda Ngom [5], Nathalie Diagne [1,3,4,*], Dioumacor Fall [1,3], Mariama Ngom [1,5,6], Diégane Diouf [7], Valerie Hocher [6], Laurent Laplaze [1], Antony Champion [8], Jill M. Farrant [9] and Sergio Svistoonoff [6,8]

[1] Laboratoire Commun de Microbiologie (LCM) Institut de Recherche pour le Développement/Institut Sénégalais de Recherches Agricoles/Université Cheikh Anta Diop, (IRD/ISRA/UCAD), Centre de Recherche de Bel Air, Dakar BP 1386, Senegal; dioumacorfall@yahoo.fr (D.F.); maringom@hotmail.fr (M.N.); laurent.laplaze@ird.fr (L.L.)

[2] Département d'Agroforesterie, Université Assane Seck de Ziguinchor, Ziguinchor BP 523, Senegal

[3] Centre National de Recherches Agronomiques (ISRA/CNRA), Bambey BP 53, Senegal

[4] Laboratoire Mixte International Adaptation des Plantes et Microorganismes Associés Aux Stress Environnementaux (LAPSE), Centre de Recherche de Bel Air, Dakar BP 1386, Senegal

[5] Département de Biologie Végétale, Université Cheikh Anta Diop de Dakar, Dakar BP 5005, Senegal; ngom_daouda@yahoo.fr

[6] Laboratoire des Symbioses Tropicales et Méditerranéennes (LSTM), (IRD/INRA/CIRAD/Université de Montpellier/Supagro), IRD TA A-82/J, Campus International de Baillarguet, 34398 Montpellier CEDEX 5, France; valerie.hocher@ird.fr (V.H.); sergio.svistoonoff@ird.fr (S.S.)

[7] UFR Environnement, Biodiversité et Développement Durable, Université du Sine Saloum El Hadj Ibrahima Niass (USSEIN), Kaolack BP 55, Senegal; diegane.diouf@ucad.edu.sn

[8] Institut de Recherche pour le Développement (IRD), Unité Mixte de Recherche DIADE (Diversité Adaptation et Développement des plantes) 911 avenue Agropolis, BP 64501, 34394 Montpellier CEDEX 5, France; antony.champion@ird.fr

[9] Department of Molecular and Cell Biology, University of Cape Town, Private Bag, Rondebosch 7701, South Africa; jill.farrant@uct.ac.za

[*] Correspondence: djighaly@yahoo.fr (P.I.D.); nathaliediagne@gmail.com (N.D.)

Received: 29 February 2020; Accepted: 24 April 2020; Published: 27 July 2020

Abstract: Land salinization is a major constraint for the practice of agriculture in the world. Considering the extent of this phenomenon, the rehabilitation of ecosystems degraded by salinization has become a priority to guarantee food security in semi-arid environments. The mechanical and chemical approaches for rehabilitating salt-affected soils being expensive, an alternative approach is to develop and utilize biological systems utilizing salt-tolerant plant species. *Casuarina* species are naturally halotolerant, but this tolerance has been shown to be improved when they are inoculated with arbuscular mycorrhizal fungi (AMF) and/or nitrogen-fixing bacteria (*Frankia*). Furthermore, *Casuarina* plantations have been proposed to promote the development of plant diversity. Thus, the aim of the current study was to evaluate the impact of a plantation comprising the species *Casuarina* inoculated with AMF and *Frankia* on the diversity of the sub-canopy and adjacent vegetation. Work was conducted on a plantation comprising *Casurina equisetifolia* and *C. glauca* variously inoculated with *Frankia* and *Rhizophagus fasciculatus* prior to field planting. The experimental area of 2500 m^2 was divided into randomized blocks and vegetation sampling was conducted below and outside of the *Casuarina* canopy in 32 m^2 plots. A total of 48 samples were taken annually over 3 years, with 24 taken from below the *Casuarina* canopy and 24 from outside the canopy. The results obtained show that co-inoculation with *Frankia* and *Rhizophagus fasciculatus* improves the height and survival rate of both species. After 4–5 years, there was greater species diversity and plant biomass in the sub-canopy environment compared with that of the adjacent environments. Our results suggest that inoculation of beneficial microbes can improve growth of *Casuarina* species and that planting of such species can improve the diversity of herbaceous vegetation in saline environments.

Keywords: *Casuarina*; salinization; diversity; rehabilitation; herbaceous vegetation

---

1. Introduction

Land salinization is characterized by high plant mortality, loss of biodiversity, and loss of soil fertility [1]. Acidic and hyper-saline halomorphic soils occupy more than 85% of the land in Palmarin (Fatick, Senegal) [2].

In recent years, an increase in salinity combined with the impacts of climate change has led to the decrease of agricultural yields. In Senegal, the area of Fatick is one of the regions most affected by salinization, and in the village of Palmarin, the agricultural production systems are threatened by the salinization of the land. Rice cultivation, which was practiced in Palmarin in the 1960s, has been abandoned because of the lack of rainfall and the increasing salinity of the land.

The increasing decline in arable land needed to provide food security for this population is exacerbated by the increasing population in this area. Official estimates taken between 1988 [3] and 2015 [4] suggest an annual growth rate of 2.72%. The lack of potential income from this area has furthermore increased the phenomenon of rural exodus, especially among the local youth [5].

It thus becomes essential to implement low-cost, ecological strategies to rehabilitate ecosystems and improve ecosystem benefits (e.g., firewood, fodder, and timber production). The use of salt-tolerant species of the Casuarinaceae family in association with symbiotic microorganisms could be a bio-remediation strategy [6,7]. Casuarinaceae form root nodules in association with the nitrogen-fixing actinobacteria *Frankia*, which enables survival in nutrient-poor soils. Furthermore, they can develop a symbiotic association with arbuscular mycorrhizal fungi (AMF) and ectomycorrhizal fungi (EM), which enhance plant water absorption and promote the uptake of nutrients, particularly, phosphorous and nitrogen [8–10]. Inoculation with salt-tolerant AMF and bacteria can improve plant salinity tolerance [11] and enhance nutrient acquisition, plant growth, and yield [12,13]. Economically, Casuarinaceae species are valued for the density of their wood and for facilitating nitrogen fertilization of soils through their symbiotic association with the soil bacteria *Frankia* [14].

What is also of importance is the regeneration of natural herbaceous vegetation in salt-affected areas, so as not to ultimately lose the natural diversity is this region. This in turn could be of reciprocal benefit. The positive effect of herbaceous vegetation on tree growth under saline conditions has been described for *Prosopis juliflora* and *Vachellia seyal* seedlings [15]. Thus, the hypothesis of our study is that salt-tolerant *Casuarina* species could have a positive effect on herbaceous diversity and reduce soil salt concentrations, thereby collectively improving growth.

To test this hypothesis, field plantations of *C. equisetifolia* and *C. glauca*, species which have been tested under controlled conditions for their ability to tolerate salt stress [7], were established in 2013. Subsequent to this, we studied (i) the effect of inoculation (*Rhizophagus fasciculatus/Frankia* strain CeD) on the growth of *C. equisetifolia* and *C. glauca* under saline environments and (ii) the impact of planting on herbaceous vegetation diversity after four years of planting in saline conditions.

Our data showed a beneficial effect of *Casuarina* on the diversity of hebaceous vegetation and that inoculation has a positive effect on wood production. We thus propose that the establishment of the vegetation layer through these programs could improve biogeochemical cycles, carbon fluxes that significantly influence the soil microbial community [16,17].

2. Materials and Methods

2.1. Field Study Site

The experimental plantation where the work was conducted is located in the Fatick region (Senegal) in the municipality of Palmarin, 14°01′14 N and 16°45′23 W (Figure 1). The plantation has a surface area of 2500 m^2 and the soil has a sandy loamy texture (analysis performed by the INP "National Institute of Pedology"). The annual rainfall is between 400 mm and 900 mm isohyets (Figure A1) and the study site is subject to periodic waterlogging during the rainy season. Soil salinity levels vary from 40 to 500 mM in the dry season. Due to its geographical location, which gives it the character of a peninsula, the temperature in the municipality of Palmarin varies between 16 °C in January and 38 °C in June.

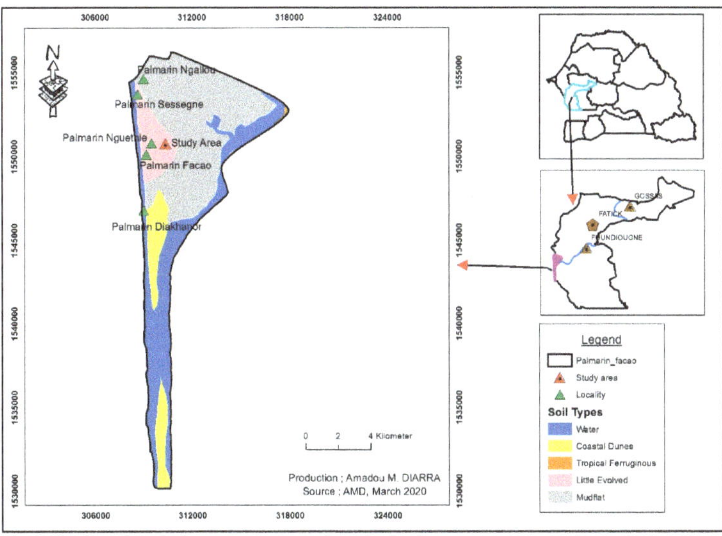

Figure 1. Map of the study site in Palmarin (Fatick, Senegal).

2.2. Experimentation in the Field

The *Casuarina* plantation was established in 2013 and is composed of *Casuarina glauca* and *Casuarina equisetifolia* which were planted according the experimental designed shown in Figure 2. Prior to field establishment, plants were inoculated with an AMF (*Rhizophagus fasciculatus* (Thaxt.) C. Walker & A. Schüßler strain DAOM227130 isolated in Quebec) according to Djighaly et al. [7] and *Frankia* (CeD) strain isolated by Diem et al. [18] from *C. equisetifolia* as described in Ngom et al. [19]. The AMF (*R. fasciculatus*) and *Frankia* strain CeD were selected on the basis of previous studies showing their ability to enhance the growth of *C. equisetifolia* and *C. glauca* under salt conditions [7,20]. The bacterial inoculum for growth of *Frankia* was prepared by culture under propionate medium (BAP) [16] for 4 days to obtain a phase of exponential bacterial growth. The plants were inoculated with 5 mL of this liquid culture (absorbance = 0.02) and the AMF inoculum containing 32 spores/g and 80% mycorrhized roots. Plants were inoculated 21 days after seeding with the following combinations: (1) plants inoculated with with *R. fasciculatus* (Rf) only, (2) plants inoculated with *Frankia* (CeD) only, and (3) plants co-inoculated with Rf and *Frankia*. Non-inoculated plants were used as controls. After 4 months in the nursery, the seedlings were transferred to the field. The experimental design included 4 blocks (each row of 8 2 × 10 subplots in Figure 2), each treatment was replicated 4 times (1 replicate per block), and in each replicate 10 plants were planted. A spacing of 2 m × 2 m between the plants was left (Figure 2). A total of 320 plants were transferred to the plot.

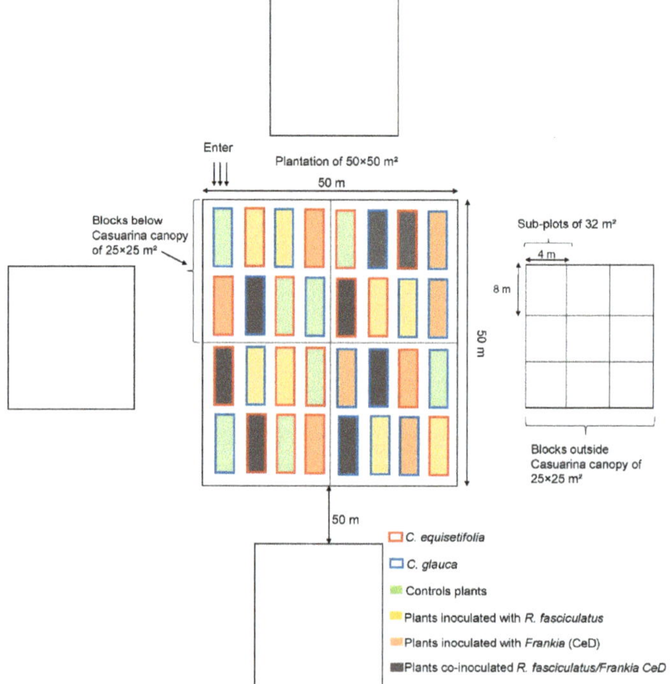

Figure 2. Experimentation design.

2.3. Physicochemical Characterization of the Site

The physical and chemical characteristics of the soil were determined at the sites before planting. Four soil samples were taken at depths of between 0–20 cm from the east, west, north, and south directions. At least 3 samples were pooled per replicate from each sampling point. Soil chemical parameters such as pH, salinity, N, C, and P were analyzed at the LAMA ("Laboratoire des Moyens Analytiques"), certified ISO 9001 version 2000.

The height and collar diameter of the plants were measured using a tape measure. This was performed after 6 months, 2 years, and 3 years to assess plant growth rates. Plant survival rate was determined by counting using the formula:

$$\text{Survival rate (\%)} = (\text{number of surviving plants})/(\text{total number of plants}) \times 100 \tag{1}$$

2.4. Diversity of Herbaceous Vegetation

An inventory of plants growing in the 2500 m^2 plantation was undertaken 3 years after planting to limit the disturbances linked to the transplanting of the *Casuarina* trees, but also to allow the planted *Casuarina* to reach a size big enough to have an impact on the fertility of the soil and the diversity of the environment. A comparison was made between the plots below the *Casuarina* canopy and outside the *Casuarina* canopy. To do this, the plantation was divided into four blocks of 25 × 25 m^2 (Figure 2). In each block, 32 m^2 sub-plots were delimited, which is the minimum area with saline land at Palmarin. The surveys were carried out using the Braun–Blanquet [20] phytosociological method. Each species inventoried was given an abundance-dominance coefficient (+ to 5), which was transformed into percentages using the method of Gillet [21].

A total of 48 vegetation surveys were carried out, including 24 below the *Casuarina* canopy, randomly distributed in each block (25 × 25 m^2) to take into account the heterogeneity of the environment. A comparison was made with 24 surveys outside the *Casuarina* canopy distributed into four blocks in eastern, western, northern, and southern directions. Botanical samples were identified in the field or in the laboratory using the flora of Senegal [22]. The species names were updated on the basis of the list of flowering plants in tropical Africa from Lebrun and Stork [23,24]. The diversity of herbaceous vegetation below and outside the *Casuarina* canopy was evaluated after 3 years (2016), 4 years (2017), and 5 years (2018) of planting.

2.5. Diversity and Regularity Indices

2.5.1. Shannon–Weaver Index

The Shannon index (H′) is used to calculate the level of species diversity in a given environment. It takes into account not only the number of species, but also the distribution of individuals within these species. It is often between 0 and 4.5, rarely higher [25]. It is calculated using the following equation:

$$H' = -\sum (ni/N) \log_2 (ni/N) \qquad (2)$$

where ni is the abundance of species i; N is the total abundance.

2.5.2. Evenness Measure

Evenness expresses the distribution of species within the association. It varies between 0 and 1, is without units, and corresponds to the ratio of Shannon index (H′) and maximum value of Shannon index (H′ max):

Evenness is represented as follows:

$$E = (H')/(H' \max) \qquad (3)$$

where H′ max = $\log_2$ (ni/N).

2.5.3. Jaccard's Similarity Index

The Jaccard similarity index allows a comparison between two sites, as it assesses the similarity between two surveys by relating the species common to both surveys to those specific to each survey. Jaccard's similarity index is represented as follows:

$$J = a/(a + b + c) \qquad (4)$$

where a is the number of species present in both samples (joint occurrences), b is the number of species present in sample one, and c is the number of species present in sample two.

2.5.4. Specific Presence Contribution (S.P.C.)

It is the ratio expressed as a percentage between the centesimal frequency of this species and the sum of the centesimal frequencies of all species; it reflects the participation of the species in covering the soil surface [26].

$$Csi\ (\%) = (Fsi/\Sigma Fsi) \times 100 \qquad (5)$$

where Fsi is the specific frequency of species i; ΣFsi is the sum of the frequency of all species; and Csi is the specific contribution of species i.

2.6. Biomass

The above-ground biomass of the sub-canopy and adjacent herbaceous vegetation was estimated within the 1 m^2 blocks by the integral harvesting method [27]. These plots were randomly distributed within the 32 m^2 plots. A total of 48 plots of 1 m^2 were used to evaluate the plant mass, 24 of which were below the *Casuarina* canopy and 24 others outside the *Casuarina* canopy.

The harvesting method consisted of harvesting the entire plant and subsequent separation into tissue types. The fresh mass of tissues was determined in the field using a spring scale. The dry mass was obtained after drying in an oven at 70 °C until a constant weight was obtained. Production is expressed in tonnes of dry biomass per hectare (t DB/ha).

Specific Contribution Biomass (Csib)

This measure reflects the participation of a species in above-ground plant biomass and is calculated as

$$Csib\ (\%) = (Wsib/\Sigma Wsib) \times 100$$

where Wsib is the weight of dry biomass of the species ib; ΣWsib is the total weight of dry biomass of all species; and Csib is the specific contribution of biomass ib [26].

2.7. Statistical Analysis

Data collected were processed using R software version 3.4.2. Normality of all data sets was assessed using the Shapiro–Wilk test and the equality of variances by the Levene test. For data following normal laws, a two-way ANOVA was performed to evaluate the effects of inoculation and species on survival rate, height, and collar diameter. Chemical parameters, coverage, specific richness, Shannon index, and dry biomass were also analyzed with Bonferonni test with a significance threshold set at 0.05. We analyzed species composition below and outside *Casuarina* canopy as a function of years using non-metric multidimensional scaling (NMDS).

3. Results

3.1. Physicochemical Characteristics of Soil

Four composite soil samples were taken at horizons of 0–20 cm before planting and 3 years after planting. The 2013 planting site had salt concentrations of approximately 0.53‰ before planting. This concentration decreased significantly 3 years after planting to 0.28‰ at the end of the dry season. The soil had an average carbon content of 2.15 g/kg, nitrogen content of 155.75 mg/kg, and phosphorus content of 42.25 before planting. This content increased to 3.3 g/kg, 275 mg/kg, and 170 mg/kg respectively for carbon, nitrogen, and phosphorus 3 years after planting (Table 1). The pH of the site was acidic; approximately 5.3 and 5.95 before planting and 3 years after planting, respectively (Table 1).

Table 1. Chemical characteristics of soil before planting in 2013 and 3 years after planting. Lower-case letters (a–b) indicate significant differences between before planting and 3 years after planting.

Analysis	pH	Salinity	Nitrogen Kjeldahl	Carbon Organic	Phosphorus Total
		‰	mg/kg	g/kg	mg/kg
Before planting	5.3 a	0.45 a	155.75 b	2.15 b	42.25 b
3 years after planting (below *Casuarina* canopy)	5.95 a	0.28 b	275 a	3.3 a	170 a

3.2. Effect of Inoculation with AMF/Frankia on the Growth of C. equisetifolia and C. glauca Planted in the More Saline Area in Palmarin

The effects of inoculation of these species in terms of survival and growth rates are given in Table 2. After 6 months of planting, in *C. equisetifolia*, all plants that were co-inoculated had survived, with those inoculated by *Frankia* or Rf only showing 80% and 77% survival, respectively. In *C. glauca*, higher survival rates occurred in plants inoculated with *Frankia* (93.34%) compared to those inoculated with Rf only (90%) or co-inoculated (87%). Two years after planting, a decrease in the survival rate was observed in both species. The highest survival rate of 67.5% was obtained in co-inoculated and *Frankia* only inoculated *C. glauca*. Compared to control plants, co-inoculation improved survival rates by 20% and 12.5% in *C. equisetifolia* and *C. glauca*, respectively (Table 2). In *C. equisetifolia*, the lowest survival rates were observed in the control plants, with 30% of the plants surviving. Of the 320 plants initially transplanted, 53% survived after 3 years (Table 2). Inoculation did not improve survival in *C. glauca* 3 years after planting. However, it did improve the survival rate of *C. equisetifolia*.

Table 2. Effect of inoculation with *R. fasciculatus* and *Frankia* on the survival rate and height of *C. equisetifolia* and *C. glauca* plants under saline stress conditions 6 months, 2 years, and 3 years after planting. Control, Rf: *R. fasciculatus*, Frankia: *Frankia* strain CeD, and Rf + *Frankia*: AMF/*Frankia* co-inoculation. Lower-case letters (a–e) indicate significant differences between control and inoculated plants for each time (6 months, 2 years, and 3 years).

	Species	**Treatments**	**Survival Rate (%)**	**Height (cm)**	**Collar Diameter (cm)**
6 months after planting	*C. equisetifolia*	Control	85 b	65.14 de	5.45 a
		Frankia	80 b	68.94 cde	6.13 a
		Rf	76.67 b	60.51 e	6.05 a
		Rf + *Frankia*	100 a	73.67 bcd	5.15 a
	C. glauca	Control	73.4 b	78.51 abc	5.75 a
		Frankia	93.34 a	79.10 abc	6.02 a
		Rf	90 a	81.98 ab	5.51 a
		Rf + *Frankia*	86.67 ab	88.00 a	6.21 a
	Species		.	***	ns
	Inoculation		**	**	ns
	Species*Inoculation		ns	ns	ns
2 years after	*C. equisetifolia*	Control	30 c	107.81 c	12.26 a
		Frankia	55 b	111.77 c	13.87 a
		Rf	47.5 b	168.13 ab	13.45 a
		Rf + *Frankia*	50 b	172.26 ab	13.49 a
	C. glauca	Control	55 b	144.40 b	11.30 a
		Frankia	67.5 a	141.29 b	12.31 a
		Rf	45 b	181.05 a	15.00 a
		Rf + *Frankia*	67.5 a	166.98 ab	12.30 a
	Species		*	*	ns
	Inoculation		**	***	ns
	Species*Inoculation		.	.	ns
3 years after	*C. equisetifolia*	Control	30 b	125.63 c	13.55 a
		Frankia	55 a	127.45 c	14.07 a
		Rf	47.5 a	180.13 ab	14.86 a
		Rf + *Frankia*	50 a	186.88 ab	14.63 a
	C. glauca	Control	55 a	152.77 bc	12.97 a
		Frankia	57 a	163.29 abc	15.33 a
		Rf	45 a	196.50 a	16.52 a
		Rf + *Frankia*	57.5 a	192.38 a	15.40 a
	Species		.	*	ns
	Inoculation		*	***	ns
	Species*Inoculation		ns	.	ns

ns, no significant difference; *, ** and *** indicate significant difference at $p < 0.1$, 0.05, 0.01, and 0.001, respectively.

In terms of the effect of inoculations on plant growth, there was no significant increase in the height of the *C. equisetifolia* plants compared to control plants 6 months after planting. However, significantly greater heights were observed in *C. glauca* that were co-inoculated compared to the control plants. After 3 years, plants of both species that were inoculated with Rf and co-inoculated had greater heights than control plants. However, inoculation did not significantly increase collar diameter compared to control plants during 3 years of planting.

3.3. Effect of the Casuarina Plantation on Species Diversity and Specific Contribution below and outside Their Canopy

The herbaceous vegetation of the site comprises 37 species from 28 genera and 14 families (Table 3). Below the *Casuarina* canopy, 24, 29, and 33 species were inventoried in 2016, 2017, and 2018, respectively. From outside the canopy, 21, 22, and 24 species were inventoried in 2016, 2017, and 2018, respectively. Of the 37 species inventoried, 14 were common in all areas. The species specifically inventoried below the *Casuarina* canopy in 2018 were *Ipomoea sinuata* Ort., *Corchorus tridens* L., and *Melochia corchorifolia* (L.).

Table 3. Presence/absence and specific contribution presence below the *Casuarina* canopy (BCC) and outside the *Casuarina* canopy (OCC).

Family	Genera	Species	BCC			OCC		
			2016	2017	2018	2016	2017	2018
	Dactyloctenium	*Dactyloctenium aegyptium* (L.) Willd	0.64	2.96	2.12	6.03	3.40	3.87
	Schisachyrium	*Schizachyrium compressa* (K. Schum.)	3.21	2.59	5.30	-	-	-
		Schizachyrium rupestre (Stapf.) Stapf	-	-	-	0.86	2.04	1.66
	Chloris	*Chloris prieurii* Kunth	3.85	3.33	3.53	3.45	3.40	5.52
		Chloris barbata (L.) Sw.	-	2.96	2.12	-	-	-
	Pennisetum	*Pennisetum polystachion* (L.) Schul.	-	3.70	3.53	2.59	4.76	4.97
	Eragrostis	*Eragrostis tenella* (L.) Beauv.	5.13	2.96	1.41	-	-	-
Poaceae		*Eragrostis tremula* Hochst. Ex Steud.	3.85	4.07	2.12	5.17	4.76	4.97
		Eragrostis aspera (Jacq.) Nees.	3.21	3.70	1.77	-	-	-
	Paspalum	*Paspalum vaginatum* (L.)	4,49	3.33	3.53	-	-	-
	Aristida	*Aristida funiculata* Trin. & Rupr.	-	2.96	3.53	5.17	5.44	1.10
		Aristida mutabilis Trin. & Rupr.	-	1.11	2.47	-	-	-
	Digitaria	*Digitaria horizontalis* Willd.	3.85	2.96	2.47	1.72	6.12	6.08
	Brachiaria	*Brachiaria lata* C.E. Hubb.	-	0.37	0.71	-	-	-
	Sporobolus	*Sporobolus robustus* Kunth	5.77	8.15	7.07	3.45	6.80	3.87
	Hibiscus	*Hibiscus rostellatus* (Guill. & Perr.)	0.64	1.48	1.41	-	-	-
		Hibiscus asper Hook.	3.21	3.33	7.42	3.45	2.72	-
Malvaceae	*Corchorus*	*Corchorus tridens* L.	-	-	0.71	-	-	-
	Sida	*Sida alba* L.	4.49	-	-	3.45	5.44	3.87
	Fuirena	*Fuirena ciliaris* (L.) Roxb.	2.56	-	-	0.86	0.68	5.52
Cyperaceae	*Cyperus*	*Cyperus esculentus* L.	1.92	4.07	2.47	-	4.76	3.31
		Cyperus bulbosus Vahl.	3.21	2.22	1.06	5.17	4.08	2.76
Amaranthaceae	*Amaranthus*	*Amaranthus greacizans* L.	0.64	3.70	3.89	6.03	2.04	2.21
	Philoxerus	*Philoxerus vermicularis* (L.) Sm.	8.33	3.33	8.13	17.24	16.33	12.71
Convolvulaceae	*Ipomoea*	*Ipomoea coptica* Willd.	1.28	0.74	1.77	5.17	2.04	3.87
		Ipomoea sinuata Ort.	-	-	3.18	-	-	-

Table 3. *Cont.*

Family	Genera	Species	BCC			OCC		
Fabaceae	Indigofera	*Indigofera berhautiana* JB. Gillett.	-	-	1.06	-	-	1.66
		Indigofera linifolia (Lf) Retz.	-	-	1.06	-	-	2.21
	Alysicarpus	*Alysicarpus ovalifolius* Schum.& Thonn.	1.92	3.33	1.77	-	-	2.21
Amaryllidaceae	Pancratium	*Pancratium trianthum* Herb.	-	1.48	3.18	-	-	-
Nyctaginaceae	Boerhavia	*Boerhavia repens* L.	3.21	4.07	3.53	3.45	3.40	3.87
Plantaginaceae	Scoparia	*Scoparia dulcis* L.	-	3.70	-	5.17	0.68	1.10
Asteraceae	Sphaeranthus	*Sphaeranthus senegalensis* DC.	7.05	4.44	1.77	2.59	5.44	3.87
Aizoaceae	Sesuvium	*Sesuvium portulacastrum* (L.) L.	14.10	8.89	3.53	7.76	9.52	7.18
Acanthaceae	Hygrophila	*Hygrophila senegalensis* (Nees) T.A	8.33	7.78	7.77	5.17	5.44	6.63
Rubiaceae	Spermacoce	*Spermacoce verticillata* L.	5.13	2.22	3.18	6.03	0.68	4.97
Sterculiaceae	Melochia	*Melochia corchorifolia* L.	-	-	0.71	-	-	-

The specific presence contribution reflects the participation of the species in the spatial occupation of the plantation. In 2016, the larger contribution towards spatial occupation below the *Casuarina* canopy was provided by the species *Sesuvium portulacastrum* (L.) L. (14.10%), *Philoxerus vermicularis* (L.) Sm. (8.33%), *Hygrophila senegalensis* (Nees) T. Anderson (8.33%), and *Sphaeranthus senegalensis* DC. (7.05%). Outside of the *Casuarina* canopy, *Philoxerus vermicularis* (17.24%), *Sesuvium portulacastrum* (7.76%), *Spermacoce verticillata* L. (6.03%), and *Amaranthus greacizans* L. (6.03%) occupied the greatest space. In 2017, below canopy domination, the species *Sesuvium portulacastrum* (L.) L. (8.89%), *Hygrophila senegalensis* (7.78%), *Sphaeranthus senegalensis* (4.44%), *Boerhavia repens* L. (4.07%), and *Cyperus esculentus* L. (4.07%) were more significant, whereas outside the *Casuarina* canopy, *Philoxerus vermicularis* (16.33%), *Sesuvium portulacastrum* (9.52%), *Melochia corchorifolia* L. (5.44%), and *Sphaeranthus senegalensis* (5.44%) were more dominant. Similarly, in 2018, greater spatial occupation below the canopy was found in the species *Philoxerus vermicularis* (8.13%), *Hygrophila senegalensis* (7.77%), *Corchorus tridens* L. (7.42%), and *Sporobolus robustus* Kunth. (7.07%), whereas outside the canopy *Philoxerus vermicularis* (12.71%), *Sesuvium portulacastrum* (7.18%), *Hygrophila senegalensis* (6.63%), and *Digitaria horizontalis* Willd. (6.08%) were more prevalant (Table 4).

Table 4. Inter-annual variation of coverage, specific richness, and diversity index below the *Casuarina* canopy (BCC) and outside the *Casuarina* canopy (OCC). Lower-case letters (a–c) indicate significant differences according to the two-way ANOVA between conditions and years.

	BCC			OCC		
Years	2016	2017	2018	2016	2017	2018
Coverage (%)	48.28 b	67.48 a	71.95 a	37.57 b	43.26 b	45.03 b
Specific richness	24 b	29 ab	33 a	21 b	22 b	24 b
Shannon index	1.9 b	2.26 a	2.39 a	1.67 c	1.76 c	2.05 b
Evenness	0.41	0.46	0.47	0.38	0.39	0.44

3.4. Coverage, Specific Richness, and Diversity Index below and outside the Casuarina Canopy

The coverage of the herbaceous vegetation below the *Casuarina* canopy was 48.28% in 2016, 67.48% in 2017, and 71.95% in 2018. Outside the canopy vegetation coverage was of 37.57% in 2016, 43.26% in 2017, and 45.03% in 2018.

Below the *Casuarina* canopy, the specific richness was 24, 29, and 33 species in 2016, 2017, and 2018, respectively. Outside of the canopy, the specific richness was 21, 22, and 24 species in 2016, 2017, and 2018, respectively (Table 5). In 2016, the below-canopy Shannon index (H') and evenness

(E) were of 1.9 and 0.41, respectively. In 2017, an increase of Shannon index and evenness of 2.26 and 0.46, respectively, was observed, whereas outside of the *Casuarina* canopy, they were 1.67 and 0.38, respectively. In 2018, a further increase of Shannon index and evenness was also observed (2.39 and 0.47) with the outside canopy indices being 2.05 and 0.44, respectively.

Table 5. Specific contribution dry biomass (Csib) below (BCC) and outside the *Casuarina* canopy (OCC).

	BCC			OCC		
	Csib			Csib		
Species	**2016**	**2017**	**2018**	**2016**	**2017**	**2018**
Dactylotenium aegyptium (L.) Willd	3.12	1.44	1.97	2.46	2.56	1.90
Schizachirium compressa (K. Schum.) Stapf	1.78	1.65	3.63	-	-	-
Schisachyrium rupestre (Stapf) Stapf	-	-	-	1.87	1.21	0.56
*Chloris prieurii*Kunt	10.79	8.21	1.14	5.32	4.56	4.64
Chloris barbata (L.) Sw.	-	10.29	5.43	-	-	-
Pennisetum polystachion (L.) Schul.	-	2.30	5.99	3.97	4.75	3.08
Eragrostis tenella (L.) Beauv.	4.24	2.01	3.41	-	-	-
Eragrostis tremula Hochst. Ex Steud.	5.53	2.73	2.68	3.77	2.41	5.84
Eragrostis aspera (Jacq.) Nees.	3.48	4.44	3.30	-	-	-
Paspalum vaginatum (L.)	1.22	5.64	2.79	-	-	-
Aristida funiculata Trin. & Rupr.	-	2.19	1.83	5.24	3.05	3.27
Aristida mutabilis Trin. & Rupr.	-	1.22	4.04	-	-	-
Digitaria horizontalis Willd.	2.59	2.48	1.43	3.10	3.84	3.70
Brachiaria lata (Schumach.) C.E. Hubb.	-	2.80	2.38	-	-	-
Sporobolus robustus Kunth.	12.97	4.40	7.80	4.45	8.07	4.21
Hibiscus rostellatus (Guill. & Perr.)	2.28	0.61	4.65	-	-	-
Hibiscus asper Hook.	-	-	0.83	-	-	-
Corchorus tridens L.	0.58	1.13	2.88	2.82	3.17	-
Sida alba L.	0.89	-	-	4.17	3.35	4.96
Fuirena ciliaris (L.) Roxb.	0.53	-	-	3.34	3.77	3.10
Cyperus esculentus L.	2.01	2.16	2.39	-	2.68	3.38
Cyperus bulbosus Vahl.	2.23	1.72	2.65	2.30	3.09	1.85
Amaranthus greacizans L.	0.94	1.69	1.20	3.89	3.77	2.68
Philoxerus vermicularis (L.) Sm.	9.31	9.59	4.36	8.42	8.82	5.46
Alysicarpus ovalifolius Schum. & Thonn.	1.35	0.74	0.43	-	-	1.42
Pancratium trianthum Herb.	-	2.17	3.80	-	-	-
Boerhavia repens L.	0.91	0.43	-	1.43	2.15	0.91
Scoparia dulcis L.	-	1.37	0.97	3.97	4.67	3.91
Sphaeranthus senegalensis DC.	2.18	1.44	-	3.02	2.26	3.11
Sesuvium portulacastrum (L.) L.	13.35	12.22	6.57	16.44	11.99	7.61
Hygrophila senegalensis (Nees)T. Anderson	14.67	9.18	6.54	12.91	10.06	7.88
Spermacoce verticillata L.	1.24	2.23	4.01	2.66	2.00	2.20
Ipomoea coptica Willd.	1.80	1.51	2.35	4.45	7.76	2.98
Ipomoea sinuata Ort.	-	-	2.39	-	-	-
Indigofera berhautiana JB. Gillett.	-	-	1.44	-	-	5.01
Indigofera linifolia (Lf) Retz.	-	-	1.23	-	-	5.01
Melochia corchorifolia L.	-	-	1.10	-	-	-

The Jaccard similarity index between below and outside the *Casuarina* canopy surveys was 0.69 in 2016. A decrease of this index to 0.59 and 0.54 in 2017 and 2018, respectively, was observed (Figure 3a). Below the *Casuarina* canopy, the Jaccard similarity index varied from 0.62 between 2016 and 2017, to 0.55 between 2016 and 2018 and 0.82 between 2017 and 2018 (Figure 3b).

Figure 3. Jaccard similarity index (**a**) between below and outside the *Casuarina* canopy surveys in 2016, 2017, and 2018 (**b**) Jaccard similarity index below *Casuarina* canopy between the years 2016–2017, 2016–2018, and 2017–2018.

NMDS was used to explore variation among species below and outside *Casuarina* canopy as a function of the years. Floristic differences were observed between the surveys conducted on below-canopy species in 2017–2018 as well as outside *Casuarina* canopy in 2016–2017 and 2018 (Figure 4).

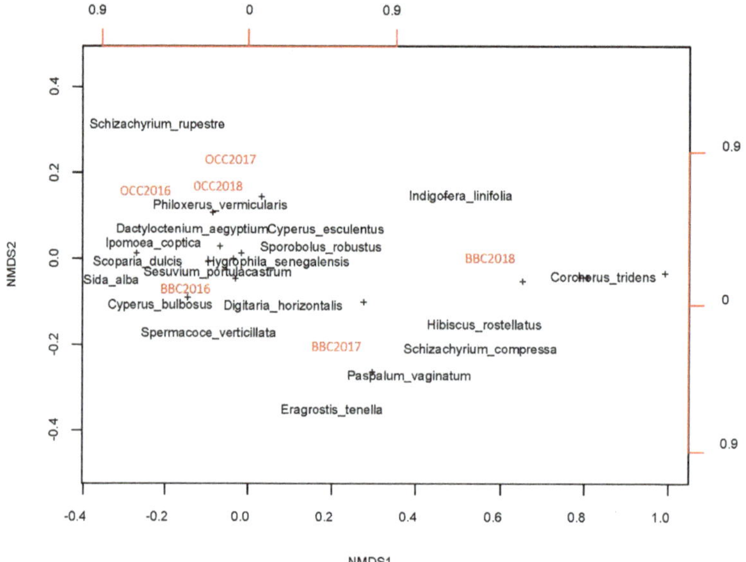

Figure 4. Non-metric multidimensional scaling (NMDS) ordination of species composition below *Casuarina* canopy (BBC) and outside *Casuarina* canopy (OCC) in 2016, 2017, and 2018.

3.5. Effect of Casuarina Species on Herbaceous Biomass in Saline Conditions

The results obtained showed a significant difference in biomass production below the *Casuarina* canopy in 2017 and 2018 (0.60 ± 0.18 t.MS/ha and 0.78 ± 0.14 t.MS/ha, respectively) compared to 2016 (0.39 ± 0.17 t.MS/ha). There is no significant difference in the biomass production outside the *Casuarina* canopy between 2016, 2017, and 2018 (Figure 5). Below the *Casuarina* canopy, biomass was significantly higher compared to outside the *Casuarina* canopy biomass in 2017 and 2018.

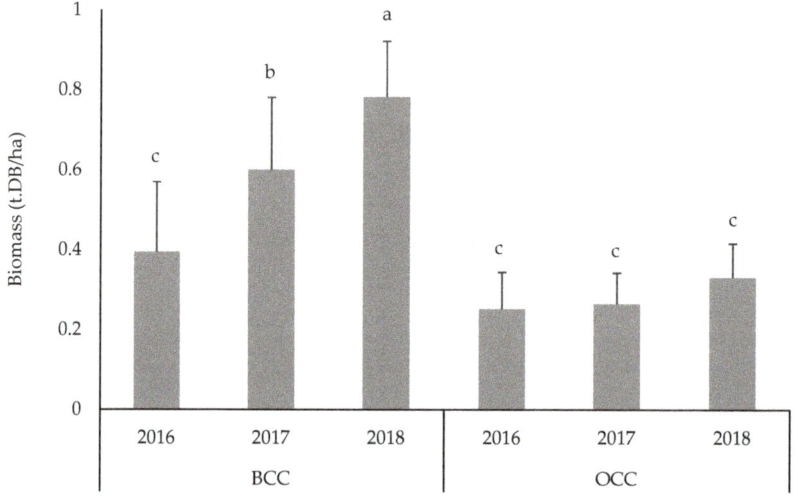

Figure 5. Inter-annual variation of biomass (BCC) below the *Casuarina* canopy (OCC) and outside the *Casuarina* canopy. Lower-case letters (a–c) indicate significant differences between the years 2016, 2017, and 2018 for each condition.

3.6. Specific Contribution of Dry Biomass below and outside the Casuarina Canopy

In 2016, the predominant species in terms of dry biomass production were *Hygrophila senegalensis* (14.64%), *Sesuvium portulacastrum* (13.35%), and *Sporobolus robustus* (12.97%). Outside the *Casuarina* canopy, the largest contributions towards dry biomass came from *Sesuvium portulacastrum* (16.44%) and *Hygrophila senegalensis* (12.91%). In 2017, the greatest biomass production was provided by below-canopy species *Sesuvium portulacastrum* (12.22%), *Chloris barbata* (10.29%), and *Hygrophila senegalensis* (9.18%). Outside the *Casuarina* canopy, the species *Sesuvium portulacastrum* (11.99%) and *Hygrophila senegalensis* (10.06%) provided the greatest dry biomass. By 2018, the largest biomass producers below the *Casuarina* canopy were *Sporobolus robustus* (7.80%), *Sesuvium portulacastrum* (6.57%), *Hygrophila senegalensis* (6.54%), and *Pennisetum polystatachion* (5.99%). Outside of the *Casuarina* canopy, biomass was greatest from *Sesuvium portulacastrum* (7.88), *Hygrophila senegalensis* (7.61%), and *Eragrostis tremula* (5.84%) (Table 5).

4. Discussion

Data provided in this study confirm that both *C. glauca* and *C. equisetifolia* can grow in saline soils. A soil is considered to be salty if it contains a minimum of 40 mM NaCl [28]. However, in our study area, the salt concentration of the soil can be 500 mM or higher during the dry season. It has been shown that *Casuarina glauca*, *Casuarina obesa*, and *Casuarina equisetifolia* will survive up to 500 mM NaCl if grown in a hydroponic medium supplemented with adequate nitrogen [19]. However, under field conditions, other factors such as drought, temperature, and the presence of other toxic ions, such as Mg^{2+}, SO_4^{2-}, HCO^{3-}, and CO_3^{2-}, can have negative effects on plant growth and survival [1]. In our study, the highest survival rates in both *C. glauca* and *C. equisetifolia* occurred in plants that had been inoculated with either *Frankia* or and co-inoculated with *Frankia* and Rf. These results show the importance of inoculation with AMF (*R. fasciculatus*) and/or nitrogen-fixing bacteria (*Frankia*) in improving *Casuarina* salt tolerance, especially in reducing post-transplant stress [7,17] and wood production. We did note, however, a decline in survival rate of *C. glauca* in the third year of this study. This could be linked to competition between the inoculum with native strains [29]. However, significant plant mortality could be related to the extreme sodium levels in the dry periods and periodic waterlogging observed at certain points in the plantation.

The effects of co-inoculation (AMF/*Frankia*) on plant growth under salt stress conditions under field conditions are poorly documented. However, similar results have been obtained under greenhouse conditions by the work of Soliman et al. [30], Ashrafi et al. [31], and Ren et al. [32] on the positive effect of co-inoculation with nitrogen-fixing bacteria and arbuscular mycorrhizal fungi on the development of *Acacia saligna*, *Medicago sativa*, and *Sesbania cannabina* plants at 250, 180, and 200 mM, respectively.

The increased height of plants co-inoculated with *Frankia* and arbuscular mycorrhizal fungi (AMF) could be explained by their facilitation in absorption of and control in levels of mineral elements. For example, selective absorption of ions, such as phosphorus, nitrogen, and magnesium and reduced absorption of Na^+ ions [33]. This selective ionic uptake reduces ionic imbalances and the antagonistic effects between ions [32] that often lead to nutritional problems caused by the saline conditions exacerbated by water deficit stress.

In terms of the effect of *C.glauca* and *C. equisetifolia* on biodiversity of herbaceous vegetation in the area, this was evident only four years (2017) after planting. There were no significant differences between the two Cassurina species with respect to the biodiversity in this vegetation (Table A1). The Shannon diversity indexes were 1.9, 2.26, and 2.39 bits in 2016, 2017, and 2018, respectively, below the *Casuarina* canopy and 1.67, 1.76, and 2.05 bits outside the *Casuarina* canopy. This inter-annual increase of the diversity could be explained by the presence of the tree, which gradually improves soil porosity, promotes good drainage [34], maintains better activity by the microbial community, and consequently increases the diversity of herbaceous vegetation. However, there is little data on the effect of trees on the diversity of herbaceous vegetation in saline conditions. Studies by Trites and Baley [35] showed a decrease in the richness of the plant community as a function of the salinity gradient and pH.

Our opposite results could be related to the heterogeneity of the electrical conductivity in the soil due to periodic waterlogging at some points of the plantation during rainy season.

Phosphorus (P) and nitrogen (N) are considered the most important elements in plant growth, and paucity thereof is has been suggested to be the greatest limitation to plant growth in most ecosystems [36]. Here, we show that the presence of *Casuarina* species results in increased phosphorus, nitrogen, and carbon contents (Table 1). These results can be explained by the decomposition of *Casuarina* litter that can improve soil fertility. The decomposition of *Casuarina* litter under semi-arid conditions results in the release of many nutrients in the following order: Ca > N > K > Mg > Na > P > Fe > Zn > Cu > Cr [37]. The Work of Hata et al. [38] suggests that decomposition of *C. equisetifolia* litter can alter the total N and N cycle in invaded forest ecosystems.

A decrease in the Jaccard similarity index between the populations below and outside *Casuarina* canopy over the successive years of this study (Figure 3) suggests differences in biological diversity among the environments. The NMDS analysis also showed a difference in the floristic composition between the surveys below *Casuarina* canopy in 2017–2018 and outside *Casuarina* canopy in 2016–2017 and 2018. This difference could be related to the beneficial effects of *Casuarina* species, which improve soil fertility through nitrogen inputs [39] and create a microclimate favorable to the establishment of herbaceous vegetation. These results can also be explained by the inter-annual presence of perennial herbaceous plants capable of improving the physicochemical properties of the soil.

The beneficial effect of herbaceous plants on decreasing soil salinity has been observed by several authors [40,41]. These herbaceous plants can improve leaching and interactions between soil chemical properties supposedly restoring soil fertility. Our results confirm this proposal with significant increases in carbon, nitrogen, and phosphorus levels occurring 3 years after planting.

A better herbaceous vegetation coverage was obtained below the *Casuarina* canopy (67.48% and 71.95% in 2017 and 2018, respectively) compared to that outside the canopy (43.26% and 45.03 in 2017 and 2018, respectively). This positive effect could be due to the presence of certain herbaceous plants capable of symbiosis with nitrogen-fixing endophytes. This has been reported for *Sporobulus robustus* with *Rhizobium* bacteria [15] and *Leptochloa fusca* with *Azoarcus* bacteria [42]. In turn, these can be considered pioneering species that improve soil fertility and encourage the establishment and growth of other herbaceous species.

The species that contributed most to the coverage of the plantation were *S. portulacastrum* and *S. robustus*, while outside the *Casuarina* canopy, *S. portulacastrum* and *Philoxerus vermicularis* were dominant. Ravindran et al. [43] have shown that *S. portulacastrum* and *S. maritima* accumulate salt in their tissues with a corresponding reduction thereof in the soil. It is estimated that these two halophytes could remove 474 and 504 kg of NaCl, respectively, per ha over a 4 month period. Our results similarly showed a decrease in soil salt concentration 3 years after establishment of the plantation. This in turn suggests that at least some of these herbaceous plants are able to reduce NaCl in the environment and make conditions more favorable for growth of *Casuarina* species by inter alia improving soil water availability. Grouzis and Akpo [44] have shown that in arid environments, the presence of trees improves biomass of other plants in their vicinity. The same could be true under saline environments. We have shown an increase in herbaceous biomass in association with the presence of *Casuarina* species, which in turn could promote the incorporation of organic biomass and facilitate salt leaching, especially Na^+ by rain.

5. Conclusions

C. equisetifolia and *C. glauca* species can be grown in saline conditions, albeit with some limitations related to the increase in salt concentration in the dry season and waterlogging conditions in the rainy seasons. Inoculation with AMF and *Frankia* increased plant growth during 3 years of planting. The *Casuarina* plantation had a positive effect on the diversity of herbaceous vegetation in salty conditions after 5 years of planting. Salt-grasses that have a greater contribution to coverage can be used to improve soil drainage and reduce salt concentration in future rehabilitation programs.

Author Contributions: P.I.D., D.N. and N.D. did the experimental work and analysis thereof and wrote the manuscript; V.H., D.F., D.D., M.N., L.L. A.C., and S.S. contributed in designing, supervision and interpretation of the results. J.M.F. edited the final manuscript. All authors have read and agreed to the published version of the manuscript.

Funding: This work was supported by the International Foundation for Sciences (IFS, no. AD/22680), the Academy of Sciences for the Developing World (TWAS, no. 11-214 RG/BIO/AF/AC I), The "Fonds d'Impulsion de la Recherche Scientifique et Technique" (FIRST) of the Ministry of Higher Education and Research of Senegal.

Acknowledgments: The "Laboratoire Mixte International Adaptation des Plantes et microorganismes associés aux Stress Environnementaux" (LAPSE). We thank Chris McMillan (International Programs Coordinator Jesuit Mission) for his participation in English correction and Amadou Mbarrick Diarra for the design of the area map. Our thanks to the whole village of Palmarin, especially Camille and Gilbert for their help in maintaining the plantation.

Conflicts of Interest: The authors declare that they have no conflict of interest.

Appendix A

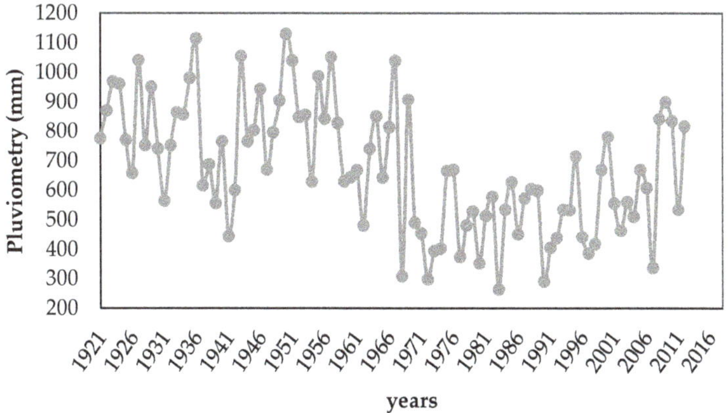

Figure A1. Rainfall data from the Fatick region (Senegal) over the period 1921–2016.

Table A1. Two-way ANOVA analysis of plant species, inoculation, and their interactions with height, collar diameter, survival rate, specific contribution presence, and dry biomass.

	Height	Collar Diameter	Survival Rate	Specific Contribution Presence	Dry Biomass
Species	**	ns	*	ns	ns
Inoculation	***	ns	**	ns	ns
Species× Inoculation	ns	ns	ns	ns	ns

ns, no significant difference; *, ** and *** indicate significant difference at $p < 0.1$, 0.05, 0.01, and 0.001, respectively.

References

1. Qadir, M.; Schubert, S. Degradation processes and nutrient constraints in sodic soils. *Land Degrad. Dev.* **2002**, *13*, 275–294. [CrossRef]
2. Diadhiou, Y.B.; Ndour, A.; Niang, I.; Niang-Fall, A. Étude comparative de l'évolution du trait de côte sur deux flèches sableuses de la Petite Côte (Sénégal): Cas de Joal et de Djiffère. *Norois* **2016**, 25–42. [CrossRef]
3. Agence National de la Statistique et de la Démographie (ANSD). Recensement Générale de la Population et de l'Habitat (RGPH). 1988. Available online: http://www.ansd.sn/ressources/rapports/Rapport_RGPH_88.pdf (accessed on 7 February 2020).
4. Agence National de la Statistique et de la Démographie (ANSD). Recensement Général de la Population et de l'Habitat, de l'Agriculture et de l'Elevage (RGPHAE). 2013. Available online: http://www.ansd.sn/ressources/rapports/Rapport-definitif-RGPHAE2013.pdf (accessed on 7 February 2020).

5. Bleibaum, F. Case Study Senegal: Environmental Degradation and Forced Migration. In *Environment, Forced Migration and Social Vulnerability*; Afifi, T., Jäger, J., Eds.; Springer: Berlin/Heidelberg, Germany, 2010. [CrossRef]

6. Potgieter, L.J.; Richardson, D.M.; Wilson, J.R.U. *Casuarina cunninghamiana* in the Western Cape, South Africa: Determinants of naturalisation and invasion, and options for management. *South Afr. J. Bot.* **2014**, *92*, 134–146. [CrossRef]

7. Djighaly, P.I.; Diagne, N.; Ngom, M.; Ngom, D.; Hocher, V.; Fall, D.; Diouf, D.; Laplaze, L.; Svistoonoff, S.; Champion, A. Selection of arbuscular mycorrhizal fungal strains to improve *Casuarina equisetifolia* L. and *Casuarina glauca* Sieb. tolerance to salinity. *Ann. For. Sci.* **2018**, *75*, 72. [CrossRef]

8. Sayed, W.F. Improving *Casuarina* growth and symbiosis with *Frankia* under different soil and environmental conditions—Review. *Folia Microbiol.* **2011**, *56*, 1–9. [CrossRef]

9. Diagne, N.; Diouf, D.; Svistoonoff, S.; Kane, A.; Noba, K.; Franche, C.; Bogusz, D.; Duponnois, R. *Casuarina* in Africa: Distribution, role and importance of arbuscular mycorrhizal, ectomycorrhizal fungi and *Frankia* on plant development. *J. Environ. Manag.* **2013**, *128*, 204–209. [CrossRef]

10. Duro, N.; Batista-Santos, P.; da Costa, M.; Maia, R.; Castro, I.V.; Ramos, M.; Ramalho, J.C.; Pawlowski, K.; Máguas, C.; Ribeiro-Barros, A. The impact of salinity on the symbiosis between *Casuarina glauca* Sieb. ex Spreng, and N_2-fixing *Frankia* bacteria based on the analysis of Nitrogen and Carbon metabolism. *Plant Soil* **2016**, *398*, 327–337. [CrossRef]

11. Hanin, M.; Ebel, C.; Ngom, M.; Laplaze, L.; Masmoudi, K. New Insights on Plant Salt Tolerance Mechanisms and Their Potential Use for Breeding. *Front. Plant Sci.* **2016**, *7*, 708. [CrossRef]

12. Abeer, H.; Abd_Allah, E.F.; Alqarawi, A.; Egamberdieva, D. Induction of salt stress tolerance in cowpea [*Vigna unguiculata* (L.) Walp.] by arbuscular mycorrhizal fungi. *Legum. Res. Int. J.* **2015**, *38*, 579–588. [CrossRef]

13. Krishnamoorthy, R.; Kim, K.; Subramanian, P.; Senthilkumar, M.; Anandham, R.; Sa, T.-M. Arbuscular mycorrhizal fungi and associated bacteria isolated from salt-affected soil enhances the tolerance of maize to salinity in coastal reclamation soil. *Agric. Ecosyst. Environ.* **2016**, *231*, 233–239. [CrossRef]

14. Zhong, C.; Zhang, Y.; Chen, Y.; Jiang, Q.; Chen, Z.; Liang, J.-F.; Pinyopusarerk, K.; Franche, C.; Bogusz, D. Casuarina research and applications in China. *Symbiosis* **2009**, *50*, 107–114. [CrossRef]

15. Fall, F.; Le Roux, C.; Bâ, A.M.; Fall, D.; Bakhoum, N.; Faye, M.N.; Kane, A.; Ndoye, I.; Diouf, D. The rhizosphere of the halophytic grass *Sporobolus robustus* Kunth hosts rhizobium genospecies that are efficient on *Prosopis juliflora* (Sw.) DC and *Vachellia seyal* (Del.) P.J.H. Hurter seedlings. *Syst. Appl. Microbiol.* **2019**, *42*, 232–239. [CrossRef] [PubMed]

16. Fierer, N.; Bradford, M.A.; Jackson, R.B. toward an ecological classification of soil bacteria. *Ecology* **2007**, *88*, 1354–1364. [CrossRef] [PubMed]

17. Goldfarb, K.C.; Karaoz, U.; Hanson, C.A.; Santee, C.A.; Bradford, M.A.; Treseder, K.; Wallenstein, M.D.; Brodie, E.L. Differential Growth Responses of Soil Bacterial Taxa to Carbon Substrates of Varying Chemical Recalcitrance. *Front. Microbiol.* **2011**, *2*, 94. [CrossRef] [PubMed]

18. Diem, H.G.; Dommergues, Y. The isolation of *Frankia* from nodules of *Casuarina*. *Can. J. Bot.* **1983**, *61*, 2822–2825. [CrossRef]

19. Ngom, M.; Gray, K.; Diagne, N.; Oshone, R.; Fardoux, J.; Gherbi, H.; Hocher, V.; Svistoonoff, S.; Laplaze, L.; Tisa, L.S.; et al. Symbiotic Performance of Diverse *Frankia* Strains on Salt-Stressed *Casuarina glauca* and *Casuarina equisetifolia* Plants. *Front. Plant Sci.* **2016**, *7*, 1. [CrossRef]

20. Braun-Blanquet, J. *Pflanzensoziologie: Grundzüge der Vegetationskunde. Mit 168 Abb. Biologische Studienbücher (Geschlossen)*; Springer: Berlin/Heidelberg, Germany, 1928; ISBN 978-3-662-02056-2.

21. Gillet, F. *La Phytosociologie Synusiale Intégrée. Guide Méthodologique*; Université de Neuchâtel, Institut de Botanique: Neuchâtel, Switzerland, 2000; 68p.

22. Berhaut, J. La Flore illustrée du Sénégal. Préf. de L. Sédar Senghor. *J. D'agriculture Traditionnelle Botanique Appliquée* **1974**, *21*, 269–270.

23. Lebrun, J.-P.; Stork, A.L. *Enumération des Plantes à fleurs d'Afrique Tropicale—Volume 1: Généralités et Annonaceae à Pandaceae*; Conservatoire et Jardin botaniques de la Ville de Genève: Geneva, Switzerland, 1991; ISBN 978-2-8277-0108-7.

24. Lebrun, J.-P.; Stork, A.L. *Enumération des Plantes à Fleurs d'Afrique Tropicale. Volume II. Chrysobalanaceae à Apiaceae*; Conservatoire et Jardin botaniques de la Ville de Genève: Geneva, Switzerland, 1992; ISBN 978-2-8277-0109-4.

25. Rougerie, G.S. Frontier et D. Pichod-Viale, Écosystèmes: Structure, fonctionnement, évolution. *Annales Géographie* **1992**, *101*, 343–344.

26. Daget, P.; Poissonet, J. Une méthode d'analyse phytosociologique des prairies. Critères d'application. *Ann. Agron.* **1972**, *22*, 5–41.

27. Rusch, G.; Armas, C.; Diouf, M.; Zapata, P.; Fall, D.; Casanoves, F.; Diémé, J.; Ibrahim, M.; Declerck, F.; Pugnaire, F. The importance of environmental gradients and tree functional attributes on tree-understory interactions in seasonally dry tropical agroforestry systems. In Proceedings of the Final Conference: "The Role of Functional Diversity for Ecosystem Services in Multi-Functional Agroforestry", Trondheim, Norway, 23–25 May 2013.

28. Munns, R.; Tester, M. Mechanisms of Salinity Tolerance. *Annu. Rev. Plant Boil.* **2008**, *59*, 651–681. [CrossRef]

29. Engelmoer, D.J.P.; Behm, J.E.; Kiers, E.T. Intense competition between arbuscular mycorrhizal mutualists in anin vitroroot microbiome negatively affects total fungal abundance. *Mol. Ecol.* **2013**, *23*, 1584–1593. [CrossRef] [PubMed]

30. Soliman, A.S.; Shanan, N.T.; Massoud, O.N.; Swelim, D.M. Improving salinity tolerance of *Acacia saligna* (Labill.) plant by arbuscular mycorrhizal fungi and *Rhizobium* inoculation. *Afr. J. Biotechnol.* **2012**, *11*, 1259–1266. [CrossRef]

31. Ashrafi, E.; Zahedi, M.; Razmjoo, J. Co-inoculations of arbuscular mycorrhizal fungi and rhizobia under salinity in alfalfa. *Soil Sci. Plant Nutr.* **2014**, *60*, 619–629. [CrossRef]

32. Ren, C.-G.; Bai, Y.-J.; Kong, C.-C.; Bian, B.; Xie, Z. Synergistic Interactions Between Salt-tolerant Rhizobia and Arbuscular Mycorrhizal Fungi on Salinity Tolerance of *Sesbania cannabina* Plants. *J. Plant Growth Regul.* **2016**, *35*, 1098–1107. [CrossRef]

33. Bar, Y.; Apelbaum, A.; Kafkafi, U.; Goren, R. Relationship between chloride and nitrate and its effect on growth and mineral composition of avocado and citrus plants. *J. Plant Nutr.* **1997**, *20*, 715–731. [CrossRef]

34. Mishra, A.; Sharma, S.D.; Khan, G.H. Rehabilitation of degraded sodic lands during a decade *of Dalbergia sissoo* plantation in Sultanpur District of Uttar Pradesh, India. *Land Degrad. Dev.* **2002**, *13*, 375–386. [CrossRef]

35. Trites, M.; Bayley, S.E. Vegetation communities in continental boreal wetlands along a salinity gradient: Implications for oil sands mining reclamation. *Aquat. Bot.* **2009**, *91*, 27–39. [CrossRef]

36. Ågren, G.I.; Wetterstedt, J.Å.M.; Billberger, M.F.K. Nutrient limitation on terrestrial plant growth—Modeling the interaction between nitrogen and phosphorus. *New Phytol.* **2012**, *194*, 953–960. [CrossRef]

37. Uma, M.; Saravanan, T.; Rajendran, K. Growth, litterfall and litter decomposition of *Casuarina equisetifolia* in a semiarid zone. *J. Trop. For. Sci.* **2014**, *26*, 125–133.

38. Hata, K.; Kato, H.; Kachi, N. Leaf litter of the invasive *Casuarina equisetifolia* decomposes at the same rate as that of native woody species on oceanic islands but releases more nitrogen. *Weed Res.* **2012**, *52*, 542–550. [CrossRef]

39. Diagne, N.; Djighaly, P.; Ngom, M.; Prodjinoto, H.; Ngom, D.; Hocher, V.; Fall, D.; Diouf, D.; Nambiar-Veetil, M.; Sy, M.; et al. Rehabilitation of Saline Lands Using Selected Salt-Tolerant Casuarina-Microorganisms Combinations. In Proceedings of the 5th International Casuarina Meeting, Chennai, India, 3–7 February 2014.

40. Dagar, J.C.; Minhas, P. (Eds.) *Agroforestry for the Management of Waterlogged Saline Soils and Poor-Quality Waters. Advances in Agroforestry*; Springer: Dordrecht, The Netherlands, 2016; ISBN 978-81-322-2657-4. [CrossRef]

41. Akhter, J.; Mahmood, K.; Malik, K.; Ahmed, S.; Murray, R. Amelioration of a saline sodic soil through cultivation of a salt-tolerant grass *Leptochloa fusca*. *Environ. Conserv.* **2003**, *30*, 168–174. [CrossRef]

42. Malik, K.A.; Bilal, R.; Mehnaz, S.; Rasul, G.; Mirza, M.; Ali, S. Association of nitrogen-fixing, plant-growth-promoting rhizobacteria (PGPR) with kallar grass and rice. *Plant Soil* **1997**, *194*, 37–44. [CrossRef]

43. Ravindranath, N.H. Mitigation and adaptation synergy in forest sector. *Mitig. Adapt. Strat. Glob. Chang.* **2007**, *12*, 843–853. [CrossRef]

44. Grouzis, M.; Akpo, L.E. Influence d'*Acacia raddiana* sur la structure et le fonctionnement de la strate herbacée dans le Ferlo sénégalais. In *Un Arbre au Désert*; Le Floc'H, É., Ed.; IRD Éditions; OpenEdition: Marseille, France, 2003; pp. 249–262.

Review

Roles of Arbuscular Mycorrhizal Fungi on Plant Growth and Performance: Importance in Biotic and Abiotic Stressed Regulation

Nathalie Diagne [1,2,*], Mariama Ngom [2,3], Pape Ibrahima Djighaly [2,3,4], Dioumacor Fall [1,2], Valérie Hocher [5] and Sergio Svistoonoff [5]

[1] Institut Sénégalais de Recherches Agricoles/Centre National de Recherches Agronomiques (ISRA/CNRA), Bambey BP 53, Senegal; dioumacorfall@yahoo.fr

[2] LMI LAPSE, Centre de Recherche de Bel Air, BP 1386, Dakar 18524, Senegal; maringom@hotmail.fr (M.N.); papadjighaly@gmail.com (P.I.D.)

[3] Laboratoire Commun de Microbiologie IRD/ISRA/UCAD, Centre de Recherche de Bel Air, Dakar 18524, Senegal

[4] Département d'Agroforesterie, Université Assane Seck de Ziguinchor, BP 523 Néma, Ziguinchor 27000, Senegal

[5] LSTM, CIRAD, INRAE, IRD, Institut Agro, TA A−82/J, Campus International De Baillarguet, University Montpellier, CEDEX 5, 34398 Montpellier, France; valerie.hocher@ird.fr (V.H.); sergio.svistoonoff@ird.fr (S.S.)

* Correspondence: nathaliediagne@gmail.com

Received: 26 June 2020; Accepted: 3 September 2020; Published: 25 September 2020

Abstract: Arbuscular mycorrhizal fungi (AMF) establish symbiotic associations with most terrestrial plants. These soil microorganisms enhance the plant's nutrient uptake by extending the root absorbing area. In return, the symbiont receives plant carbohydrates for the completion of its life cycle. AMF also helps plants to cope with biotic and abiotic stresses such as salinity, drought, extreme temperature, heavy metal, diseases, and pathogens. For abiotic stresses, the mechanisms of adaptation of AMF to these stresses are generally linked to increased hydromineral nutrition, ion selectivity, gene regulation, production of osmolytes, and the synthesis of phytohormones and antioxidants. Regarding the biotic stresses, AMF are involved in pathogen resistance including competition for colonization sites and improvement of the plant's defense system. Furthermore, AMF have a positive impact on ecosystems. They improve the quality of soil aggregation, drive the structure of plant and bacteria communities, and enhance ecosystem stability. Thus, a plant colonized by AMF will use more of these adaptation mechanisms compared to a plant without mycorrhizae. In this review, we present the contribution of AMF on plant growth and performance in stressed environments.

Keywords: AM fungi; plant; abiotic stress; biotic stress

1. Introduction

Arbuscular mycorrhizal fungi (AMF) are soil microorganisms that form a symbiotic relationship with 80–90% of vascular plant species and 90% of agricultural plants [1], including most agricultural crops, particularly cereals, vegetables, and horticultural plants. They have a ubiquitous distribution in global ecosystems that are primarily defined by the global distribution of known plant hosts [2,3]. AMF are classified as a member of subkingdom *Mucoromyceta* and the phylum *Glomeromycota* including three classes (*Glomeromycetes*, *Archaeosporomycetes*, and *Paraglomeromycetes* [4]). AMF belong to 11 families, 25 genera, and nearly 250 species [5,6]. *Glomeromycota* are obligate symbionts that rely on the carbon substrates provided by their host plants (up to 20% of plant-fixed carbon) to survive [7,8]. In return, the fungi improve the supply of water and nutrients, such as phosphate and nitrogen to the host plant through extraradical and intraradical hyphae, arbuscules, and the root apoplast interface [9].

Based on fossil records and molecular data, this symbiosis dates back to the first appearance of land plants, about 400 to 450 million years ago [1]. The Arbuscular mycorrhizal (AM) symbiosis is probably the most widespread beneficial interaction between plants and microorganisms [9]. Several studies have reported that they play a crucial role in plant nutrition and growth in stressed conditions and enhance a number of essential ecosystem processes [7,10].

The purpose of this review is to summarize knowledge about AM associations, in particular the beneficial effects on host plants and soil. First, the role of AMF in plant nutrition, growth, and resistance to biotic and abiotic stress is considered. The indirect contribution of AMF in soil aggregation and stability is discussed. Finally, the diversity of interactions between AMF and other soil microorganisms are examined (Figure S1).

2. Contribution of Arbuscular Mycorrhizal Fungi to Plant Nutrition and Growth

Among beneficial microbes, AMF are one of the most widespread symbiotic fungi colonizing the majority of agricultural plants [11]. The effects of AMF on plant growth and physiological elements contents have been widely studied in many species including relevant crops such as *Solanum lycopersicum* L. [12,13], *Sorghum bicolor* (L.) Moench [10,14], *Withania somnifera* (L.) Dunal [15], *Cucurbita maxima* Duchesne [16], *Piper longum* L. [17], *Phaseolus vulgaris* L. [18], *Panicum hemitomon* Schult [19], and some free fruits such as *Citrullus lanatus* (Thunb.) Matsum. & Nakai [20], *Musa acuminata* Colla [21], and *Prunus cerasifera* L. [22]. In all these species, AMF improved plant growth parameters [10,12–15,17] and the uptake of several major nutrients such as nitrogen and phosphorus in stressed conditions [23,24]. This growth stimulation is linked to the fact that AMF extends the absorbing network beyond the nutrient depletion zones of the rhizosphere, which allows access to a larger volume of soil [25]. Furthermore, fungal hyphae are much thinner than roots and are able to penetrate smaller pores and uptake more nutrients [26].

By extending the root absorbing area, AMF increases the total absorption surface of inoculated plants and thus improves plant access to nutrients, particularly those whose ionic forms have a poor mobility rate or those which are present in low concentration in the soil solution [1]. It is calculated that the rate of water transport from external hyphae to the root ranges from 0.1 [27] to 0.76 μL H_2O h^{-1} per hyphal infection point [28]. Furthermore, AMF contributes approximately 20% to total plant water uptake [29], highlighting the role of the symbiosis in the water status of host plants. AMF significantly improved *Cucurbita maxima* growth and metabolism, such as the concentrations of fat, crude protein, crude fiber, and carbohydrates in shoot and root systems of inoculated plants compared to control treatment [16]. Inoculation with this fungus significantly increased plant growth as well as phytochemical constituents such as sugar, protein, phenol, tannin, and flavonoid content [15]. In watermelon (*Citrullus lunatus* Thunb.), mycorrhizal colonization was found to improve not only the plant yield and water use efficiency but also the quality of the fruits [30]. Similar results were obtained in mycorrhizal tomato plants with an increase in the concentrations of sugars, organic acids, and vitamin C in fruits [12]. It has been demonstrated by [31] that AMF improved peach seedlings' performance under the potted conditions, and also significantly elevated K, Mg, Fe, and Zn concentrations in leaves and roots, Ca concentration in leaves, Cu and Mn concentrations in roots, which were obviously dependent on the AMF species. Compared to three AMF ([*Funneliformis mosseae* (T.H. Nicolson & Gerd.) C. Walker & A. Schüßler 2010, *Glomus versiforme* (P. Karst.) S.M. Berch 1983, and *Paraglomus occultum* (C. Walker) J.B. Morton & D. Redecker 2001), *F. mosseae* exhibited the best mycorrhizal efficiency on growth and nutrient acquisition of peach seedlings [31]. Compared to uninoculated plants, AMF inoculation had positive effects on the growth of carrot and sorghum [14]. In carrot, *Scutellospora heterogama* (T.H. Nicolson & Gerd.) C. Walker & F.E. Sanders 1986, *Acaulospora longula* Spain & N.C. Schenck 1984, and *F. mosseae* had a positive effect on the growth of the host, whereas AMF had only weak effects on the growth of red pepper and leek [14].

Therefore, it is important to mention that the extent to which a host plant benefits vary with the AMF species used [14]; and macro and micro-nutrients uptake could depend partly not only on the fungal partner but also on the host plant [32]. A study carried out by [33] indicates that the contribution of the mycorrhizal pathway to nutrient acquisition also depends on fungal effects on the activity of the plant pathway and on the efficiency with which both partners interact and exchange nutrients across the mycorrhizal interface [34]. For various crops such as sweet potato [35] or pepper plant [17], the beneficial effect on plant nutrient content has also been shown to be dependent on fungal diversity.

A similar positive effect was reported in sorghum with an enhancement of plant height, the number of leaves, biomass, total nitrogen, phophorus and potassium uptake [10]. Although, among some species of native AMF tested (*Glomus aggregatum* N.C. Schenck & G.S. Sm 1982, *F. mosseae*, *Acaulospora longula*, and *Acaulospora scrobiculata* Trappe 1977) some species like *Acaulospora scrobiculata* are more efficient for improvement of all these parameters in sorghum [10]. The effect of an AMF, *F. mosseae* was examined regarding the morphological and biochemical properties of different genotypes of the medicinal plant *W. somnifera*, commonly called Ashwagandha [15]. In addition, several studies reported that the responses of plants to colonization by AMF vary depending on inoculum composition, and a combination of mycorrhizal fungi is more effective than a monospecific inoculum [10,14,17,18].

AMF colonization by *F. mosseae* or *R. intraradices* (N.C. Schenck & G.S. Sm.) C. Walker & A. Schüßler 2010) increased both the survival and growth (by over 100%) of micropropagated transplants of *Prunus cerasifera* L., compared with either uninoculated controls or transplants inoculated with the ericoid mycorrhizal species *Hymenoscyphus ericae* (D.J. Read) Korf & Kernan 1983 [22]. Thus, inoculation of woody species' seedlings under nursery conditions is a valuable strategy to produce seedlings with good vigor, which would translate into high survival and growth at the field [36–38].

AMF play an important role in biofortification [39,40]. AMF inoculation may affect selenium uptake from soil and the level of antioxidant compounds in vegetable crops such as the green asparagus *Asparagus officinalis* L. Research carried out by [41,42] showed increasing selenium (Se) content in wheat grain through inoculation. It has been found by [43] that AMF modifies the concentration and distribution of nutrients within wheat and barley grain. Inoculation with AMF improves the grain nutritional content in protein, Fe, and Zn [44]. Under distinct environmental conditions, [45] concluded that AMF symbiosis positively affected the Zn concentration in various crop plant tissues. AMF can contribute substantially to the Zn nutrition of cereal crops such as bread wheat and barley but the role played by AMF on Zn uptake depends on the functional compatibility between AMF isolate and inoculated cereal species [40].

It is well-known that AMF symbiosis specifically induces the expression of transporters such as the plant aquaporin (AQ) genes, Pi transporters (PT), ammonium transporter (AMT), nitrate transporter (NT), sulfur (S) transporter, Zn transporter, carbon transporter, protein transporter etc. [46–50]. In wheat plants treated with *F. mosseae* and *R. intraradices* Zn concentration is 1.13–2.76 times higher than non-inoculated plants observed [51]. Further, it has been demonstrated that fungal form a network called mycorrhizal networks (MNs) that improve nutrients transfer between plants through the extension of fungal mycelium [52,53]. Also called common mycorrhizal networks, these MNs can integrate multiple plant species and multiple fungal species that interact, provide feedback, and adapt, which comprise a complex adaptive social network [52]. Results obtained by [54] confirm the role of AMF in driving biological interactions among neighboring plants.

3. Role of Arbuscular Mycorrhizal Fungi in Alleviation of Abiotic Stresses in Plants

AMF respond differently to abiotic stresses such as drought, flooding, extreme temperatures, salinity, and heavy metals [55–58]. Studies carried out by [59] showed that AMF communities would change in composition in response to abiotic stress [59]. These stresses reduce AMF diversity and alter AMF community composition resulting in an AMF community with a higher proportion of species that are phenotypically similar, because they are more tolerant of that specific abiotic stress. Changes in the diversity of AMF will feed back into the plant community and cause corresponding changes in the diversity of plant species and productivity [60]. The feedback will become stronger with climate change [61]. AMF will adapt to abiotic stress independently of its host plant. However, despite the negative effects of stresses on AMF, several studies have demonstrated that AMF symbiosis improves plant growth, hydration, and physiology under various environmental stress conditions like salinity, drought, and the presence of heavy metals [62,63]. Depending on the stress, the benefits of AMF to plant partners can vary [64], and mycorrhizal fungal types vary in their responses to climate change.

3.1. AMF and Plant Drought Tolerance

Drought is one of the major stresses that can reduce plant productivity considerably [11]. Water constraints provoke stomatal closure with a subsequent reduction of CO_2 influx resulting in a decrease in photosynthetic activity and carbon partitioning [65] and a decrease in plant productivity and agricultural yield. It has been demonstrated that AMF improves plant performance in drought stress [66]. Table 1 shows the different effects of AMF on plants' drought tolerance. Mycorrhizal plants deal with water deficit through drought mitigation and drought tolerance [67]. A drought mitigation strategy is mediated by indirect AMF benefits and enhanced water uptake, whereas drought tolerance includes a combination of direct AMF benefits that improve the plant's innate ability to cope with the stress [11].

Through drought mitigation, the improvement of plant fitness by AMF is possibly due to the increased surface area for water absorption provided by AMF hyphae [68], increased access to small soil pores, or improved apoplastic water flow. Numerous studies have related the mechanisms of drought mitigation [69–72]. It has been shown by [73] that in field-grown tomato plants, root colonization by the AMF *R. intraradices* enabled plants to grow well under water stress conditions through an improvement of nutrient contents and water use efficiency. These beneficial effects of AMF on tomato tolerance to water stress have also been reported in several other plant species such as *Lactuca sativa* L. [74], *Triticum aestivum* L. [75], *Lavandula spica* L. [76], *Allium cepa* L. [77], *Trifolium repens* L. [78], *Pistacia vera* L. [79], *Acacia auriculiformis* A.Cunn. ex Benth, *Albizia lebbeck* (L.) Benth., *Gliricidia sepium* (Jacq.) Kunth ex Walp. And *Leucaena leucocephala* (Lam.) de Wit [80].

Other mechanisms are involved in plant response to drought stress, among them, the production of phytohormones. Hormone homeostasis regulates plant tolerance against abiotic stresses. Abscisic acid (ABA) is the most fundamental stress hormonal signal, modulating transpiration rate, root hydraulic conductivity, and aquaporin expression. ABA responses regulate stomatal conductance and other related physiological processes [81]. ABA induces stomatal closure and reduces cell water loss. Inoculation with AMF influences the control of stomata functioning by the regulation of abscisic acid [81,82]. A lower ABA concentration was found in roots and leaves of mycorrhizal plants versus nonmycorrhizal plants under drought stress [10,71,72]. It has been also demonstrated that Jasmonic acid (JA) interacts with abscisic acid to regulate plant responses to water stress conditions [82]. JA can mitigate water stress in plants [83]. This hormone is involved in the regulation of the expression and the abundance of aquaporins and plays an important role in water uptake and transport, on stomate and root hydraulic conductance.

Other mechanisms are involved in plant tolerance to drought. Among these mechanisms, the osmotic adjustment that allows plants to maintain their turgor and physiological activity [11] by accumulating compatible solute compounds such as sugars, proline, glycine betaine, polyamines, and organic acids such as oxalate and malate. As described in saline conditions, drought stress induces the production of reactive oxygen species (ROS) [84].

Other phytohormones, such as strigolactone and auxin, are involved in plant water stress regulation [85]. It has been demonstrated that the inoculation with AMF strengthens strigolactone and auxin responses to drought stress [74].

Table 1. Contribution of AMF in helping plants to cope with drought stress.

Host Plants	AMF Strains	Responses Related to AMF Inoculation	References
Zea mays L. *Solanum lycopersicum* L.	*Rhizophagus irregularis* (Błaszk., Wubet, Renker & Buscot) C. Walker & A. Schüßler 2010	Enhanced apoplastic water flow	[69]
S. lycopersicum L.	*F. mosseae* *R. irregularis*	Increased plant height and biomass, intrinsic water use efficiency (iWUE) index, stomatal density, capacity to absorb CO_2 and proline concentrations, and reduced hydrogen peroxide, leaf and root ABA contents	[71]
Lycopersicon. esculentum L.	*Glomus clarum* T.H. Nicolson & N.C. Schenck 1979	Improved leaf area, dry mass, stomatal conductance, photosynthetic activity, and root hydraulic conductivity	[72]
L. esculatum L.	*R. irregularis*	Increased plant height, number of primary branches, flowers, and fruits, shoot and root dry matter, N and P contents, fruit yields, leaf relative water content (RWC), water use efficiency (WUE) and quality of fruits (less acidity and quantities of ascorbic acid and total soluble solids)	[73]
S. lycopersicum *Lactuca sativa* Linn.	*R. irregularis*	Improved shoot dry weight, stomatal conductance, photosystem II efficiency, ABA and strigolactone contents	[74]
Triticum aestivum L.	*R. fasciculatus* *F. mosseae*	Enhanced stomatal conductance and leaf osmotic adjustment	[75]
Lavandula spica L.	*R. irregularis* *F. mossea*	Increased biomass, N, K and water contents, and reduced antioxidant compounds (glutathione, ascorbate and H_2O_2)	[76]
Allium cepa L.	*Glomus etunicatus* W.N. Becker & Gerd. 1977	Improved fresh and dry weights and phosphorus nutrition	[77]
Trifolium repens L.	*R. irregularis*	Enhanced dry weight, nutrients content (P, K, Ca, Mg, Zn and B), relative water content, proline concentrations, and glutathione reductase activity	[78]
Pistacia vera L.	*G. etunicatum*	Increased shoot and root weights, leaf area, total chlorophyll, and flavonoids contents, nutrient concentrations (P, N, K, Ca, Fe, Zn, and Cu), soluble sugar, proline, and soluble proteins contents, CAT and POD activities	[79]
L. esculatum L. *Capsicum annuum* L.	*Rhizophagus irregularis* *Rhizophagus fasciculatus* (Thaxt.) C. Walker & A. Schüßler, 2010	Improved biomass, root length, shoot length, and chlorophyll contents, and reduced proline concentration	[86]

3.2. AMF and Plant Flooding Tolerance

Some AMF can also face other constraints, such as flooding [87]. In wetland ecosystems, [88] found a higher diversity of the communities of AMF. In these conditions, AMF could greatly improve the growth of plants through enhanced absorption of nutrient elements [89] (Table 2). These authors found that a considerable amount of P was transported to rice plants via the mycorrhizal pathway under wetland conditions. Results obtained by [90] indicate that AMF may assist *Phragmites australis* (Cav.) Trin. ex Steud. in coping with a medium frequency of drying-rewetting cycles. Similar results were found by [91], who showed the contributions to the flood tolerance of *Pterocarpus officinalis* Jacq. seedlings by improving plant growth and P acquisition in leaves. It has been demonstrated by [92], that the better growth of plants with mycorrhiza on flooding condition is linked to the improvement of osmotic adjustment. Since AMF needs oxygen to thrive, flooding may inhibit AMF colonization, and accordingly previous studies have found a decrease in the degree of AMF colonization with flooding along wetland gradients [92]. Furthermore, results obtained by [93] have indicated that the distribution of AMF in tropical low flooding forest is related to the characteristics of vegetation, chemical parameters of the soil and identity of the AMF.

Table 2. Contribution of AMF in helping plants to cope with flooding stress.

Host Plants	AMF Strains	Responses Related to AMF Inoculation	References
Panicum hemitomon Schult. Schultes *Leersia hexandra* Schwartz	*Acaulospora trappei* R.N. Ames & Linderman 1976, *Scutellospora heterogama, A Acaulospora laevis* Gerd. & Trappe 1974, *Glomus leptotichum* N.C. Schenck & G.S. Sm 1982, *G. etunicatum* and Glomus *gerdemannii* S.L. Rose, B.A. Daniels & Trappe 1979	Improve phosphorus (P) nutrition Greater tissue P concentrations	[19]
Pterocarpus officinalis (Jacq.)	*Glomus intraradices*	Improve plant growth and P acquisition in leaves	[91]
Aster tripolium L.	*Glomus geosporum* T.H. Nicolson & Gerd.) C. Walker 1982	Higher concentrations of soluble sugars and proline	[93]

3.3. AMF and Plant Tolerance to Extreme Temperatures

Temperature is one of the most important environmental stresses that can negatively affect the growth and productivity of plants [94]. It is well known that AMF improves plant performance to tolerate temperature (heat and cold) [95] stress by enhancing water and nutrient uptake, improving photosynthetic capacity and efficiency, protecting plants against oxidative damage, and increasing the accumulation of osmolytes [96] (Table 3). At low or high temperatures, it was reported that shoot and dry root weights of mycorrhizal plants were higher than non-mycorrhized plants [97]. At high temperature, AMF help plants to develop their root system for absorption of water to ensure high photosynthetic capacity and to prevent the photosynthetic apparatus from being damaged [98]. Regarding cold stress, [99] showed that *G. versiforme* was often more effective than *R. irregularis* for the alleviation of low-temperature stress in the winter and the spring cultivars, whereas *R. irregularis* was more effective in increasing the survival rate. For these authors, the response to cold stress depends on the AMF strains. At low-temperature conditions, the inoculation of Barley (*Hordeum vulgare* L.) with AMF resulted in improved growth, photosynthesis, osmotic homeostasis and potassium uptake [99]. However, extremely low and high temperatures reduced AMF fungal growth inhibited the formation of the extra radical hyphal network and AMF fungal activity. AMF may, therefore, play a key role in the mitigation of climate change [100], such as tolerance to a wider range of temperatures. AMF also improves the reduction of N2O emissions by enhancing N uptake and assimilation by plants. Consequently, as a result, the soluble N in the soil decreases and it can negatively affect the denitrification process [100,101]. However, seasonal and climate changes, including temperature fluctuations, can affect the temporal structuring of AMF communities [102].

Table 3. Contribution of AMF in helping plants to cope with extreme temperatures stress.

Host Plants	AMF Strains	Responses Related to AMF Inoculation	References
Cucumis sativus L.	*R. irregularis*	Increases the photosynthetic efficiency of cold-stressed cucumber seedlings by protecting their photosynthetic apparatus against light-induced damage and increasing their carbon sink.	[51]
Zea mays L.	*Funneliformis (Glomus) species.*	Regulated photosystem (PS) II heterogeneity	[98]
Hordeum vulgare L.	*G. versiforme* *R. irregularis*	Increasing the survival rate, alleviation of low-temperature stress	[99]
Cyclamen persicum Mill.	*R. fasciculatum*	Enhanced biomass production and heat stress response Increase activity of antioxidative enzymes such as superoxide dismutase and ascorbate peroxidase	[103]
Elymus nutans Griseb.	*F. mosseae*	Less oxidative damage Promoted plant growth and enhanced the level of chlorophyll and antioxidant compounds such as glutathione and soluble sugars	[104]

3.4. AMF and Plant Tolerance to Salinity

AMF have been known to occur naturally in saline environments [105,106]. Their contribution to the improvement of the growth of several plant species under saline conditions is well known [107,108] (Table 4). This is mainly related to a combination of biochemical, physiological, and nutritional effects [97,109–116]. Among the mechanisms involved in salinity tolerance in AMF inoculated plants, we have the enhancement of water absorption capacity and nutrient uptake, the accumulation of osmoregulators like proline and sugars [105], the ionic homeostasis [87,117], and the reduction in Na^+ and Cl^- uptake [118]. In addition, it has been demonstrated that AMF colonization improves stomatal conductance and reduces the oxidative damage in plants exposed to salinity [112,119,120]. For example, inoculation with *F. mosseae* of tomato plants irrigated with saline water significantly increased plant biomass, fruit fresh yield, and shoot contents of P, K, Cu, Fe, and Zn [121]. In another study, plant root colonization with this same AMF decreased Na concentration and enhanced the activity of various enzymes related to the mitigation of salt stress [97]. Similar results were reported for cereals such as wheat [112] and maize [119]. Under salt-stressed conditions, [116] showed that AMF inoculation significantly reduced the oxidative damage in wheat plants. They also reported higher gas exchange capacity, stomatal conductance, and concentrations of sugars, free amino acids, proline, and glycinebetaine in plants colonized by AMF [116]. *Zea mays* plants inoculated separately with three native AMF showed better biomass production and higher shoot potassium and proline contents, as compared to non-mycorrhizal plants [119]. It has also been demonstrated that AMF alleviates the deleterious effects of salt on plants growth in Acacia species [109,122,123]. This is mainly related to greater nutrient acquisition, total chlorophylls, carbohydrates, and proline contents elevated K/Na ratios in root and shoot tissues, and changes in root morphology [109,124] in mycorrhizal plants as compared nonmycorrhizal plants. These results were more significant when salt-stressed plants were co-inoculated with selected rhizobium strains, ectomycorrhizal fungi and/or endophytic bacteria, in addition to AMF [86,123,125]. Nevertheless, it should be mentioned that in studies where different strains were tested, the extent of AMF response on plant growth as well as root colonization varied with fungal species, and with the level of salinity [126]. These authors found that AMF differ in their effects on *Chrysanthemum morifolium* Ramat. plants under saline conditions and *Diversispora versiformis* (P. Karst.) was the most active than *F. mosseae*. Regarding ions homeostasis, AMF improve ionic balance by filtering effect of AMF structures both in the soil and in the root that prevents the entry of toxic Na^+ ions [107,124].

Table 4. Contribution of AMF in helping plants to cope with salinity stress.

Host Plants	AMF Strains	Responses Related to AMF Inoculation	References
Lycopersicon esculentum L.	*F. mosseae*	Increased plant growth, fruit weight, and yield, chlorophyll content, concentrations of P and K, antioxidant enzymes activities (SOD, CAT, POD, and APX), and reduced Na concentration in leaves	[97]
Capsicum annuum L.	*R. irregularis*	Increased leaf area, mineral content, proline, sugars, and cell membrane integrity, and reduced shoot content of Na	[106]
Acacia auriculiformis A. Cunn. ex Benth.	*R. fasciculatus* *Glomus macrocarpum* Tul. & C. Tul. 1845	Increased root and shoot weights, and greater nutrient acquisition, changes in root morphology, and electrical conductivity of the soil	[109]
Solanum lycopersicum L.	*R. irregularis*	Enhanced shoot and root dry weights, chlorophyll and proline concentrations, nutrient uptake (P, Ca, and K), stomatal conductance, the activity of ROS scavenging enzymes (APX, CAT, POD, and SOD) and protecting photochemical processes of PSII	[111]
Triticum aestivum L.	*G. etunicatum* *F. mosseae R. irregularis*	Increased plant growth, nutrient uptake and grain yield, and reduced concentrations of Na^+ and Cl^-	[112]
Zea mays L.	*R. irregularis* (isolate EEZ 58) *R. irregularis* (Ri CdG) *S. constrictum* (Trappe) Sieverd., G.A. Silva & Oehl 2011 (Sc CdG) *Claroideoglomus etunicatum* (W.N. Becker & Gerd.) C. Walker & A. Schüßler, 2010 (Ce CdG) (Ce CdG)	Improved K^+ and Na^+ homeostasis, shoot and root dry weights, K concentration in shoots, and reduced Cl and Na contents in shoots	[119]
Digitaria eriantha Steud.	*R. irregularis*	Increased stomatal conductance, antioxidant enzymes activities (CAT et APX), jasmonate content, and reduced root and shoot hydrogen peroxide accumulation	[120]
Lycopersicon esculentum Mill. Cv. Marriha	*F. mosseae*	Improved plant biomass, fruit fresh yield and shoot contents of P, K, Cu, Fe and Zn, and reduced shoot Na concentrations	[121]
Acacia nilotica Willd.	*R. fasciculatum*	Improved root and shoot biomass, and nutrient concentrations (P, Zn, K, and Cu), and Na concentration	[122]
Acacia saligna (Labill.) H.L. Wendl.	AMF	Improved plant growth and dry weight, nodulation parameters, chlorophylls, carbohydrates, proline and nutrient contents (N, P, K and Ca) and reduced Na concentrations	[123]
Acacia auriculiformis *Acacia mangium*	*R. irregularis*	Enhanced plant growth and nodulation, and nutrient contents (P, N)	[125]
Gossypium arboreum L.	*F. mosseae* (isolate GM1) *F. mosseae* (isolate GM2)	Increased biomass and phophorus concentrations	[127]
Cucumis sativus L.	*C. etunicatum* *R. irregularis* *F. mosseae*	Enhancing the biomass, synthesis of pigments, activity of antioxidant enzymes, including superoxide dismutase, catalase, ascorbate peroxidase, and glutathione reductase, and the content of ascorbic acid Enhancing jasmonic acid, salicylic acid and several important mineral elements (K, Ca, Mg, Zn, Fe, Mn, and Cu) Reducing the uptake of deleterious ions like Na^+	[128]

Table 4. *Cont.*

Host Plants	AMF Strains	Responses Related to AMF Inoculation	References
Glycine max L. Merrill	*C. etunicatum* *R. irregularis* *F. mosseae*	Protected soybean genotypes from salt-induced membrane damage Reduced the production of hydrogen peroxide and lipid peroxidation Improved plant growth and symbiotic performance by stimulating the endogenous level of auxins that contribute to improved root systems and nutrient acquisition under salt stress	[129]

3.5. AMF and Plant Tolerance to Heavy Metals

Mining sites and polluted sites with heavy metal contain AMF that are specifically adapted to soil pollution by heavy metals [130,131].

Numerous studies showed that more than 80% of surveyed plants growing on mining sites are colonized by AMF, and a great number of AMF species and a large AMF diversity exist in various mining-impacted sites [131]. These authors summarized studies that showed that AMF exhibit significant positive effects, such as increased plant survival, enhanced growth and nutrition, improved soil structure and quality, and greater plant re-establishment.

Several studies revealed that mycorrhiza could be used as a stress-reducing agent in soils contaminated by heavy metals helping plants to survive in such stressed conditions [24,41,86,132–135] (Table 5). Heavy metal remediation by AMF can happen through hyphal "metal binding", which reduces the bioavailability of elements, such as Cu, Pb, Co, Cd, and Zn [136]. The alleviation of heavy metal toxicity by AMF depends on the fungal partner, plant growth conditions, the type of heavy metal, and its concentration [130]. Inoculation with AMF showed the best results in terms of percentage of seed germination, the sustainability of seedlings, fresh weight, and dry weight of plants. In two different heavy metal-polluted soils, root colonization of maize plants with Glomus isolates reduced heavy metal concentrations in shoots and roots, and increased the contents of essential elements like K, P, and Mg in roots [133]. This result was more significant in maize plants colonized with the Glomus isolate Br1 from *Viola calaminaria* (DC.) Lej. (zinc violet) compared with plants grown with a common Glomus strain or to non-colonized controls. These authors also reported distinct differences in the cellular distribution of heavy metals and essential elements in mycorrhiza compared with the non-colonized control roots, suggesting that AMF might cope with heavy metal toxicity for each metal individually [133]. When maize was grown in soil contaminated with Cd, AMF inoculation significantly increased growth and reduced Cd uptake, suggesting that AMF can be used in association with plants for the mitigation of heavy metal such as Cd in soils [137]. AMF expressed some metal transporters that play a crucial role in heavy metal regulation. In recent years, several Zn transporters were identified in AMF, such as GintZnT1 from *R. irregularis* [138]. Several putative genes coding for Cu, Fe, and Zn transporters have been also identified [139]. These transporters could be involved in heavy metal tolerance in plants inoculated by AMF.

Table 5. Contribution of AMF in helping plants to cope with heavy metals stress.

Host Plants	AMF Strains	Responses Related to AMF Inoculation	References
Trigonella foenum-graecum L.	*F. mosseae*	Better plant performance	[132]
Zea mays L.	*Glomus* isolates	Improved dry weight and contents of essential elements (K, P, and Mg), and distinct differences in the cellular distribution of heavy metals and essential elements	[133]
Lonicera japonica Thunb.	*G. versiforme R. intraradices*	Decreased Cd concentrations in shoots and roots, Reduced Cd concentrations in shoots but increased Cd concentrations in roots	[140]

Table 5. *Cont.*

Host Plants	AMF Strains	Responses Related to AMF Inoculation	References
Solanum lycopersicum L.	*F. mosseae* (syn. *Glomus mosseae*), *R. intraradices* (syn. *Glomus intraradices*) *C. etunicatum* (syn. *Glomus etunicatum*)	AMF reduced the production of malonaldehyde and hydrogen peroxide by mitigating oxidative stress. AMF strengthened the plant's defense system and provide efficient protection against Cd stress	[141]
Populus alba Villafranca *Populus nigra* Jean Pourtet	*F. mosseae* or *R. irregularis*	Alleviation of Cu and Zn phytotoxicity	[142]
Trifolium pratense L.	*Glomus mosseae*	Decreases in Zn uptake and in root and shoot concentrations	[143]

4. Role of AM Fungi in Alleviation of Biotic Stresses in Plants

The role of microsymbionts in the management of biotic stresses is gaining importance. Numerous studies have proven that AMF reduces the damage caused by various plant pathogens [144–146] (Table 6). For example, studies carried out by [147] demonstrate that the severity of charcoal root-rot disease in soybean can be reduced by AMF inoculation. In the presence of *Fusarium*, arbuscular colonization increased shoot dry weight [144]. Colonization by AMF has a protective effect termed mycorrhiza-induced resistance (MIR) [146,148,149], which provides systemic protection against a wide range of attackers and shares characteristics with systemic acquired resistance (SAR) after pathogen infection and induced systemic resistance (ISR) following root colonization by non-pathogenic rhizobacteria [150]. This mycorrhiza-induced resistance to diseases was described by [150]. These authors showed that AMF increases the production of antioxidant enzymes in plants, which can act as a defense against pathogens and other stresses. In addition to the activation of plant defense mechanisms, several other reasons for reduced damage of pathogens by AMF have been reported such as improved nutrient status of the host plant, change in root growth and morphology, competition for colonization sites and host photosynthates, and microbial changes in the mycorrhizosphere [144,151,152]. The improvement of plant growth may have a positive effect because mycorrhizae can facilitate the regrowth of tissues after attacks. However, it may have negative effects because as plant nutrition improves, it becomes more nutritive or attractive to herbivore insects [153]. The contribution of AMF in the protection of plants against pathogenic fungi and nematodes is well documented. Although, it should be mentioned that the effectiveness of the interactions varies depending on the host plant and the cultural system [5]. The colonization of tomato roots by *F. mosseae* induced systemic resistance against both the sedentary nematode *Meloidogyne incognita* (Kofoid & White, 1919) and the migratory nematode *Pratylenchus penetrans* (Cobb, 1917) Filipjev & Schuurmans Stekhoven, 1941 [144]. The presence of this AMF reduced nematode infection by 45% and 87% for *M. incognita* and *P. penetrans*, respectively, in AMF-colonized plants as compared to controls. Further studies carried out on root exudates have shown that the reduction of nematode infection in mycorrhizal plants is probably related to an alteration of their root exudates by AMF [154]. Indeed, the application of mycorrhizal root exudates further reduced nematode penetration in mycorrhizal plants and temporarily paralyzed nematodes, in comparison with the application of water or non-mycorrhizal root exudates. Root colonization by *F. mosseae* caused also a reduction in galling, nematode reproduction and morphometric parameters of females in tomato plants inoculated with *Meloidogyne javanica* (Treub, 1885) [151]. Regarding pathogenic fungi, mycorrhizal inoculation with *F. mosseae* significantly alleviated tomato diseases caused by *Alternaria solani* (Ellis & G. Martin) L.R. Jones & Grout 1896 and *Fusarium oxyspoum Fusarium oxysporum* Schltdl. 1824, respectively [155–157]. This beneficial effect was more pronounced when plants were inoculated with AMF and sprayed with hormonal inducers (Jasmonic acid and Salicylic acid), suggesting a synergistic and cooperative effect between them leading to an enhanced induction and regulation of disease resistance [156]. These beneficial effects of AMF have also been reported in potato [158] and chickpea [152] infected with pathogens.

In contrast to the well-known effect on pathogenic nematodes and fungi, there are relatively few studies on the impact of AMF on herbivore insects [159,160]. The meta-analysis conducted by [159] on the published and unpublished studies on this topic revealed that the effects of mycorrhizal fungi on herbivore insects varied depending on the parameter measured and the degree of herbivore feeding specialization. Some AMF like *R. intraradices* tended to have a negative effect on chewer performance, as opposite to other fungal species studied [159]. For example, in *Plantago lanceolate* L., mycorrhizal infection increased the resistance of leaves to the chewing insect *Arctia caja* (Linnaeus, 1758), while the performance of the sucking insect *Mysus persicae* (Sulzer) was greater on mycorrhizal plants [161]. It has been shown by [160] that parasitism of *Chromatomyia syngenesiae* by *Diglyphus isaea* (Walker, 1838) was lower on mycorrhizal plants, while in the laboratory the effects of three species of AMF on parasitism rates were dependent on the species of AMF [160].

Among biotic constraints that mostly affect productivity in developing countries is the African witchweed Striga (hereafter, referred to as "Striga"), mainly in sub-Saharan Africa. This parasitic plant is a socioeconomic problem that has forced many poor farmers to abandon their farms due to high infestation rates [162]. Soil microorganisms including AMF can inhibit or suppress Striga germination [163]. AMF can affect the interaction between Striga and cereals [164]. These authors showed that AMF negatively impacted Striga seed germination, reduced the number of Striga seedlings attaching and emerging, and delayed the emergence time of Striga. Studies carried out by [165] confirmed the effectiveness of AMF in protecting sugarcane against Striga infestation as well as promoting crop growth and reducing the soil Striga seed bank. By doing so, AMF enhanced the performance of the plant host, allowing it to better withstand Striga damage [164].

Table 6. Contribution of AMF in helping plants to cope with several biotic stresses.

Host Plants	Disease or Pathogen	AMF Strains	Responses Related to AMF Inoculation	References
Solanum lycopersicum L.	*M. incognita*	*F. mosseae*	Induced systemic resistance against both the sedentary nematode *Meloidogyne incognita* and the migratory nematode *Pratylenchus penetrans*	[144]
L. esculentum	*Fusarium oxysporum f. sp. lycopersici*	*Glomus sp.*	Production of antimicrobial compounds from the mycorrhizal root that arrested the mycelial growth of the fungal pathogen Reduced the disease incidence Increased the plant growth, dry weight, N, P, K content, chlorophyll content and yield of the plant	[145]
Glycine max (L.) Merr.	*Macrophomina phaseolina* (Tassi) Goid 1947	*R. irregularis*	Improve plant height and the number of functional leaves	[147]
Solanum lycopersicum L.	*M. javanica*	*F. mosseae*	Caused also a reduction in galling, nematode reproduction and morphometric parameters of females in tomato plants inoculated with	[151]
	Alternaria solani Sorauer 1896 *Fusarium oxysporum*	*F. mosseae*	Alleviated tomato diseases	[155,156]
Saccharum officinarum L.	*Striga hermonthica* Del Benth 1836	*G. etunicatum, Scutellospora fulgida* Koske & C. Walker 1986, *G. margarita*	Stimulated plant growth, plant biomass and physiological parameters of plants in the presence of Striga	[165]
Solanum lycopersicum L.	*Cladosporium fulvum* Cooke 1883	*F. mosseae*	Higher resistance against subsequent pathogen infection higher fresh and dry weight increases in total chlorophyll contents and net photosynthesis rate	[166]
Astragalus adsurgens var. Shanxi Yulin	*Erysiphe pisi* DC 1805	*C. etunicatum, G. versiforme, F. mosseae*	Increased the shoot and root growth of standing milkvetch even though their presence in the roots increased susceptibility to powdery mildew.	[167]

Table 6. *Cont.*

Host Plants	Disease or Pathogen	AMF Strains	Responses Related to AMF Inoculation	References
Capsicum annum	*Pythium aphanidermatum* (Edson) Fitzp 1923	*Glomus sp.*	Mycelial growth of the fungal pathogen reduced the disease incidence and increased the growth and yield of crop plants	[168]
Cicer arietinum L.	*Fusarium wilt*	*Glomus hoi* S.M. Berch & Trappe 1985, *R. fasciculatum*	Increased total contents of P and N in treated plants	[169]
Cucumis melo L.	*Fusarium wilt*	*F. mosseae*	Greatest capacity for reduction of disease incidence	[170]
Arachis hypogaea L.	*Sclerotium rolfsii* Sacc 1911	*R. fasciculatum, Gigaspora margarita, Cucumis melo* L., *A. laevis*, and *Sclerocystis dussii* (Pat.) Höhn. 1910	Eliminated the damaging effects of *S. rolfsii*	[171]
Solanum melongena L. *Cucumis sativus* L.	*Verticillium dahliae* Kleb 1913 *Pseudomonas lacrymans* (Smith and Bryan) Carsner 1970	*G. versiforme*	Alleviated wilt symptoms caused by *V. dahliae*	[172]
S. tuberosum	*Potato virus* Y (PVY)	*R. irregularis*	Milder symptoms and significant stimulation of shoot growth were observed in PVY-infected plants inoculated	[173]
Nicotiana tabacum L.	*Tobacco mosaic virus* (TMV) *Cucumber green mottle mosaic virus* (CGMMV)	*R. irregularis*	Showed reduced disease symptoms and virus titer if compared to non-mycorrhizal plants	[174]
Zea mays L.	*Striga hermonthica* Del Benth 1836	*G. etunicatum, Scutellospora fulgida* Koske & C. Walker 1986, *G. margarita*	Reduced *Striga* plant incidence, plant biomass, and phosphate content	[175]
Sorghum bicolor (L.) Moench	*S. hermonthica*	*F. mosseae*	Improved the performance of sorghum	[176]

5. Interaction between AMF and Other Beneficial Soil Microorganisms

AMF interacts with a wide assortment of soil microorganisms [177–179]. Interactions can be either positive, neutral, or negative on the mycorrhizal association or on other microorganisms in the rhizosphere [180–183]. They may be involved in nutrient acquisition, biological control of root pathogens, improvement of plant tolerance to abiotic stress and soil quality [184].

5.1. Interaction between AMF and Nitrogen Fixing Bacteria

5.1.1. Interaction between AMF and Rhizobia

AMF and nitrogen-fixing bacteria provide plants with essential soil nutrients, and the expectation is that co-inoculation will result in the strongest synergistic effects. Several experiments have demonstrated a positive effect of the interactions between AMF and nodulating rhizobial bacteria [185]. Numerous studies showed a beneficial effect of the interaction between AMF and Rhizobia in legumes such as *Amorpha canescens* Nutt., *Lens culinaris* Medik. [186,187], *Glycine max* [188,189], *Pisum sativum* L. [190], *Vicia faba* L. [191] and *Lathyrus sativus* L. [192]. These interactions can induce changes in the microbial environment through their secretions [193]. It has been demonstrated by [193] that the dual inoculation with Rhizobium and AMF is more effective for promoting the growth of Faba bean in alkaline soils than the individual treatment. The dual inoculation of AMF and nitrogen-fixing bacteria increased nodulation, nitrogen fixation, plant growth and yield. Co-inoculation with selected AMF and rhizobia also improved outplanting performance, plant survival, and biomass development of woody legumes in desertified ecosystems [194]. This is in accordance with the strong synergistic effects of AMF and rhizobia inoculation found on the biomass production of *Atriplex canescens* (Pursh) Nutt. [186]. However, these interactions were contingent on several factors, like the amounts of phosphorus and nitrogen available [186]. Similar results were obtained by [188], which showed a synergistic relationship dependent on N and P status between rhizobia and AM fungi on soybean growth.

5.1.2. Interaction between AMF and Frankia

As was observed in legumes, the synergistic effects of AMF and the nitrogen fixing actinobacteria Frankia improve actinorhizal plants performance in various environments [195–201]. In *Casuarina cunninghamiana* Miq. and *C. equisetifolia* L., dual inoculation with a mycorrhizal fungus and a Frankia isolate significantly increased the height of seedlings and trees, depending on the levels of available phosphorus [199]. Regarding the impact on nitrogen levels, [200] studies on the natural abundance of 15N in four species of Casuarina revealed that the interactions between Frankia and AMF were species-dependent and is also influenced by the availability of P and N. This is in accordance with [198] studies which showed that AMF plus Frankia had no effect in wood volume growth of *C. cunninghamiana*, while this parameter was favored by fertilization with N. Interactions between AMF and Frankia have been well studied in the pioneer species, *Alnus glutinosa* L. Gaertn. [195–197]. In a highly alkaline anthropogenic sediment, dual inoculation of Black alder plants with *R. intraradices* and Frankia spp significantly increased leaf area, shoot height, total biomass, and N and P leaf contents when compared with the uninoculated control, the Frankia spp. and the *R. intraradices* treatments alone [197]. In addition, the numbers and dry weight of root nodules, as well as the development of the AM symbiosis were greater when dual inoculation was performed, suggesting a synergistic effect of these mycrosymbionts, which allowed *A. glutinosa* plants to grow under these hostile conditions. Although, it has been reported in this species that interactions between AMF and Frankia are not always positive [195]. Indeed, in a glasshouse experiments, early interactions between different AM species (*Glomus hoi, F. mosseae, Gigaspora rosea* T.H. Nicolson & N.C. Schenck 1979, *A. scrobiculata* and *Scutellospora castanea* C. Walker 1993) and Frankia lead to a depressive effect on plant biomass [195]. This effect resulting possibly from the competition with microsymbionts for resources such as photosynthates may only be temporary.

5.2. Interaction between AMF and Plant Growth Promoting Rhizobacteria

AMF also interact with the plant growth-promoting Rhizobacteria (PGPR) [202–205]. PGPR are soil and rhizosphere bacteria that can be of benefit to plant growth by several different mechanisms such as a symbiotic N2 fixation, ammonia production, solubilization of mineral phosphate, and other essential nutrients, production of plant hormones, and control of phytopathogenic microorganisms [206]. Under field conditions, dual inoculation of *Schizolobium parahyba* (Vell.) S.F. Blake, 1919 with AMF and PGPR increased wood yield by about 20% compared to the application of chemical fertilizers alone [199]. Inoculation of *Acacia gerrardii* Benth. with AMF and *Bacillus subtilis* Ehrenberg, 1835 Cohn, 1872 induced a significantly greater shoot and root dry weight, nodule number and leghemoglobin content than those inoculated with AMF or *B. subtilis* alone under salt stress [207]. These authors found a positive synergistic interaction between AMF and *B. subtilis* regarding nitrate and nitrite reductase and nitrogenase activities and the contents in total lipids, phenols, fiber, and osmoprotectants such as glycine, betaine, and proline [207]. AMF and phosphate solubilizing bacteria (PSB) could interact synergistically because PSB solubilizes sparingly available phosphorous compounds into orthophosphate that AMF can absorb and transport to the host plant [179].

Under drought stress, dual inoculation with AMF and PGPR allowed to alleviate water deficit damage and improve water stress tolerance in *Cupressus arizonica* Greene through a better accumulation of ascorbate peroxidase and glutathione peroxidase, in comparison to plants inoculated with a single microorganism [203]. Inoculation with AMF and PGPR also has a positive effect on plant metabolism. [203] showed that co-inoculation of *Stevia rebaudiana* Bertoni) Bertoni, with AMF and PGPR enhanced significantly plant growth parameters, NPK, total chlorophyll, and stevioside contents.

5.3. Benefits of the Tripartite Symbiosis (AMF, Nitrogen Fixing Bacteria, PGPR, or Ectomycorrhizal Fungi)

The benefits of the tripartite symbiosis (AMF, nitrogen-fixing bacteria, PGPR, or ectomycorrhizal fungi) are also known [208–210]. As shown by [209], inoculation with a combination of AMF, Frankia, Azospirillum, and Phosphobacterium significantly increased the total height and total biomass of *C. equisetifolia* plants. The same authors measured an enhanced nutrient uptake for N, P, K, Ca, and Mg in triple inoculated plants. AMF can also co-exist with ectomycorrhizal fungi (EMF) and improves plant growth [208,211]. EMF and AMF plant colonization do not occur simultaneously. Generally, AMF is established first, followed by EMF, but less often, EMF establishes first then reduces AMF colonization by forming a mantle that acts as a barrier to AMF infection. However, when AMF is established first, it has no negative effects on EMF infection [211]. In different *Acacia crassicarpa* A. Cunn. ex Benth. provenances, combined EMF and AMF symbioses improved plant growth and the rhizobial nodulation process [208]. Similar results were obtained in *Robinia pseudoacacia* L. plants inoculated with a combination of EMF, AMF, and Rhizobium [212]. In *C. equisetifolia* plants, [213] observed that inoculation with both AMF and EMF significantly increased biomass and P content compared to plants inoculated with either AMF or EMF alone. Inoculation of *C. equisetifolia* with Frankia increased nitrogen fixation, ectomycorrhizal, and endomycorrhizal colonization [214]. However, an antagonistic effect was observed when both symbionts were inoculated and were generally the result of high ectomycorrhizal colonization [211].

5.4. Interaction between AMF and Mycorrhization Helper Bacteria

AMF interacts positively with Mycorrhization Helper Bacteria (MHB) [215–217]. Because of the beneficial effect of bacteria on mycorrhizae, the concept of Mycorrhization Helper Bacteria (MHB) was created. Five possible ways of action of MHB on mycorrhiza were proposed by [217]: in the receptivity of the root to the mycobiont, in root-fungus recognition, in fungal growth, in the modification of the rhizospheric soil, and in the germination of fungal propagules. These authors showed that MHB is fungus-specific but not plant-specific.

6. Conclusions

AMF play an important role in improving the adaptation to biotic and abiotic plant stresses and to alleviate the effects of these stress on plants. Their role in increasing plant growth and yield, disease resistance, biotic and abiotic tolerance provides an environmentally friendly solution to reduce the use of hazardous pesticides and industrial fertilizers. However, more research is needed to test in the field the results obtained in the laboratory and in the greenhouse. The application of this knowledge in real environments and according to biogeographical zones becomes essential in order to promote their industrial production for a large scale used and increase their impact to ensure enough food for every human being on the planet now and in the future. As an ecofriendly method, some work must be done by researchers, private and public sectors, to promote the use of AMF by increasing their production, particularly in developing countries where AMF inocula are not accessible and not affordable.

Supplementary Materials: The following are available online at http://www.mdpi.com/1424-2818/12/10/370/s1, Figure S1: The role of AMF on plant growth in stressed environment.

Author Contributions: N.D., M.N. and P.I.D. wrote the manuscript; S.S., V.H. and D.F. corrected the paper and supervised the work. All authors have read and agreed to the published version of the manuscript.

Funding: The "Fonds d'Impulsion de la Recherche Scientifique et Technique" (FIRST) of the Ministry of Higher Education, Research and Innovation of Senegal.

Acknowledgments: The authors thank Richard Moise Alansou DIEME for helping in design the supplementary material of this paper.

Conflicts of Interest: The authors declare that they have no conflict of interest.

References

1. Smith, S.E.; Read, D.J. *Mycorrhizal Symbiosis*; Academic Press: Cambridge, MA, USA, 2010; ISBN 978-0-08-055934-6.
2. Kivlin, S.N.; Hawkes, C.V.; Treseder, K.K. Global diversity and distribution of arbuscular mycorrhizal fungi. *Soil Biol. Biochem.* **2011**, *43*, 2294–2303. [CrossRef]
3. Wang, B.; Qiu, Y.-L. Phylogenetic distribution and evolution of mycorrhizas in land plants. *Mycorrhiza* **2006**, *16*, 299–363. [CrossRef] [PubMed]
4. Tedersoo, L.; Sánchez-Ramírez, S.; Kõljalg, U.; Bahram, M.; Döring, M.; Schigel, D.S.; May, T.; Ryberg, M.; Abarenkov, K. High-level classification of the Fungi and a tool for evolutionary ecological analyses. *Fungal Divers.* **2018**, *90*, 135–159. [CrossRef]
5. Schüßler, A.; Schwarzott, D.; Walker, C. A new fungal phylum, the Glomeromycota: Phylogeny and evolution. *Mycol. Res.* **2001**, *105*, 1413–1421. [CrossRef]
6. Spatafora, J.W.; Chang, Y.; Benny, G.L.; Lazarus, K.; Smith, M.E.; Berbee, M.L.; Bonito, G.; Corradi, N.; Grigoriev, I.V.; Gryganskyi, A.; et al. A phylum-level phylogenetic classification of zygomycete fungi based on genome-scale data. *Mycologia* **2016**, *108*, 1028–1046. [CrossRef] [PubMed]
7. Siddiqui, Z.A.; Pichtel, J. Mycorrhizae: An Overview. In *Mycorrhizae: Sustainable Agriculture and Forestry*; Springer Science and Business Media LLC: Berlin, Germany, 2008; pp. 1–35.
8. Johns, C.D. Agricultural Application of Mycorrhizal Fungi to Increase Crop Yields, Promote Soil Health and Combat Climate Change. Future Directions International. 2014. Available online: https://www.futuredirections.org.au/publication/agricultural-application-of-mycorrhizal-fungi-to-increase-crop-yields-promote-soil-health-and-combat-climate-change/ (accessed on 11 August 2020).
9. Parniske, M. Arbuscular mycorrhiza: The mother of plant root endosymbioses. *Nat. Rev. Genet.* **2008**, *6*, 763–775. [CrossRef]
10. Nakmee, P.S.; Techapinyawat, S.; Ngamprasit, S. Comparative potentials of native arbuscular mycorrhizal fungi to improve nutrient uptake and biomass of *Sorghum bicolor* Linn. *Agric. Nat. Resour.* **2016**, *50*, 173–178. [CrossRef]
11. Posta, K.; Duc, N.H. Benefits of Arbuscular Mycorrhizal Fungi Application to Crop Production under Water Scarcity. *Drought Detect. Solut.* **2020**. Available online: https://www.intechopen.com/books/drought-detection-and-solutions/benefits-of-arbuscular-mycorrhizal-fungi-application-to-crop-production-under-water-scarcity (accessed on 11 August 2020). [CrossRef]
12. Bona, E.; Cantamessa, S.; Massa, N.; Manassero, P.; Marsano, F.; Copetta, A.; Lingua, G.; D'Agostino, G.; Gamalero, E.; Berta, G. Arbuscular mycorrhizal fungi and plant growth-promoting pseudomonads improve yield, quality and nutritional value of tomato: A field study. *Mycorrhiza* **2016**, *27*, 1–11. [CrossRef]
13. Gamalero, E.; Trotta, A.; Massa, N.; Copetta, A.; Martinotti, M.G.; Berta, G. Impact of two fluorescent pseudomonads and an arbuscular mycorrhizal fungus on tomato plant growth, root architecture and P acquisition. *Mycorrhiza* **2003**, *14*, 185–192. [CrossRef]
14. Kim, S.J.; Eo, J.-K.; Lee, E.-H.; Park, H.; Eom, A.-H. Effects of Arbuscular Mycorrhizal Fungi and Soil Conditions on Crop Plant Growth. *Mycobiology* **2017**, *45*, 20–24. [CrossRef] [PubMed]
15. Parihar, P.; Bora, M. Effect of mycorrhiza (*Glomus mosseae*) on morphological and biochemical properties of Ashwagandha (*Withania somnifera*) (L.) Dunal. *J. Appl. Nat. Sci.* **2018**, *10*, 1115–1123. [CrossRef]
16. Al-Hmoud, G.; Al-Momany, A. Effect of Four Mycorrhizal Products on Squash Plant Growth and its Effect on Physiological Plant Elements. *Adv. Crop. Sci. Technol.* **2017**, *5*, 1–6. [CrossRef]
17. Gogoi, P. Differential effect of some arbuscular mycorrhizal fungi on growth of *Piper longum* L. (Piperaceae). *Indian J. Sci. Technol.* **2011**, *4*, 119–125. [CrossRef]
18. Ibijbijen, J.; Urquiaga, S.; Ismaili, M.; Alves, B.J.R.; Boddey, R.M. Effect of arbuscular mycorrhizal fungi on growth, mineral nutrition and nitrogen fixation of three varieties of common beans (*Phaseolus vulgaris*). *New Phytol.* **1996**, *134*, 353–360. [CrossRef]
19. Miller, S.P.; Sharitz, R.R. Manipulation of flooding and arbuscular mycorrhiza formation influences growth and nutrition of two semiaquatic grass species. *Funct. Ecol.* **2000**, *14*, 738–748. [CrossRef]

20. Ban, D.; Ban, S.G.; Oplanić, M.; Horvat, J.; Novak, B.; Žanić, K.; Žnidarčič, D. Growth and Yield Response of Watermelon to in-row Plant Spacings and Mycorrhiza. *Chil. J. Agric. Res.* **2011**, *71*, 497–502. [CrossRef]

21. Rodríguez-Romero, A.S.; Guerra, M.S.P.; Jaizme-Vega, M.D.C. Effect of arbuscular mycorrhizal fungi and rhizobacteria on banana growth and nutrition. *Agron. Sustain. Dev.* **2005**, *25*, 395–399. [CrossRef]

22. Berta, G.; Trotta, A.; Fusconi, A.; Hooker, J.E.; Munro, M.; Atkinson, D.; Giovannetti, M.; Morini, S.; Fortuna, P.; Tisserant, B.; et al. Arbuscular mycorrhizal induced changes to plant growth and root system morphology in *Prunus cerasifera*. *Tree Physiol.* **1995**, *15*, 281–293. [CrossRef]

23. Jansa, J.; Forczek, S.T.; Rozmoš, M.; Püschel, D.; Bukovská, P.; Hršelová, H. Arbuscular mycorrhiza and soil organic nitrogen: Network of players and interactions. *Chem. Biol. Technol. Agric.* **2019**, *6*, 10. [CrossRef]

24. Song, Z.; Bi, Y.; Zhang, J.; Gong, Y.; Yang, H. Arbuscular mycorrhizal fungi promote the growth of plants in the mining associated clay. *Sci. Rep.* **2020**, *10*, 1–9. [CrossRef] [PubMed]

25. Smith, F.A.; Smith, F.A. Roles of Arbuscular Mycorrhizas in Plant Nutrition and Growth: New Paradigms from Cellular to Ecosystem Scales. *Annu. Rev. Plant. Biol.* **2011**, *62*, 227–250. [CrossRef] [PubMed]

26. Allen, M.F. Linking water and nutrients through the vadose zone: A fungal interface between the soil and plant systems. *J. Arid. Land* **2011**, *3*, 155–163. [CrossRef]

27. Allen, M.F. Influence of vesicular-arbuscular mycorrhizae on water movement through *Bouteloua gracilis* (H.B.K.) lag ex steud. *New Phytol.* **1982**, *91*, 191–196. [CrossRef]

28. Faber, B.A.; Zasoski, R.J.; Munns, D.N.; Shackel, K. A method for measuring hyphal nutrient and water uptake in mycorrhizal plants. *Can. J. Bot.* **1991**, *69*, 87–94. [CrossRef]

29. Ruth, B.; Khalvati, M.; Schmidhalter, U. Quantification of mycorrhizal water uptake via high-resolution on-line water content sensors. *Plant. Soil* **2011**, *342*, 459–468. [CrossRef]

30. Kaya, C.; Higgs, D.; Kirnak, H.; Taş, I. Mycorrhizal colonisation improves fruit yield and water use efficiency in watermelon (*Citrullus lanatus* Thunb.) grown under well-watered and water-stressed conditions. *Plant. Soil* **2003**, *253*, 287–292. [CrossRef]

31. Wu, Q.-S.; Li, G.-H.; Zou, Y.N. Roles of arbuscular mycorrhizal fungi on growth and nutrient acquisition of peach (*Prunus persica* L. Batsch) seedlings. *J. Anim. Plant. Sci.* **2011**, *21*, 746–750.

32. Trouvelot, S.; Bonneau, L.; Redecker, D.; Van Tuinen, D.; Adrian, M.; Wipf, D. Arbuscular mycorrhiza symbiosis in viticulture: A review. *Agron. Sustain. Dev.* **2015**, *35*, 1449–1467. [CrossRef]

33. Ravnskov, S.; Jakobsen, I. Functional compatibility in arbuscular mycorrhizas measured as hyphal P transport to the plant. *New Phytol.* **1995**, *129*, 611–618. [CrossRef]

34. Farmer, M.; Li, X.; Feng, G.; Zhao, B.; Chatagnier, O.; Gianinazzi, S.; Gianinazzi-Pearson, V.; Van Tuinen, D. Molecular monitoring of field-inoculated AMF to evaluate persistence in sweet potato crops in China. *Appl. Soil Ecol.* **2007**, *35*, 599–609. [CrossRef]

35. Jansa, J.; Smith, F.A.; Smith, S.E. Are there benefits of simultaneous root colonization by different arbuscular mycorrhizal fungi? *New Phytol.* **2008**, *177*, 779–789. [CrossRef] [PubMed]

36. Zangaro, W.; Nisizaki, S.M.A.; Domingos, J.C.B.; Nakano, E.M. Mycorrhizal response and successional status in 80 woody species from south Brazil. *J. Trop. Ecol.* **2003**, *19*, 315–324. [CrossRef]

37. Vandresen, J.; Nishidate, F.R.; Torezan, J.M.D.; Zangara, W. Inoculação de fungos micorrízicos arbusculares e adubação na formação e pós-transplante de mudas de cinco espécies arbóreas nativas do sul do Brasil. *Acta Bot. Bras.* **2007**, *21*, 753–765. [CrossRef]

38. Tahat, M.; Kamaruzaman, S.; Radziah, O.; Kadir, J.; Masdek, H. Plant Host Selectivity for Multiplication of *Glomus mosseae* Spore. *Int. J. Bot.* **2008**, *4*, 466–470. [CrossRef]

39. Golubkina, N.A.; Krivenkov, L.; Sękara, A.; Vasileva, V.; Tallarita, A.; Caruso, G. Prospects of Arbuscular Mycorrhizal Fungi Utilization in Production of Allium Plants. *Plants* **2020**, *9*, 279. [CrossRef]

40. Coccina, A.; Cavagnaro, T.R.; Pellegrino, E.E.; Ercoli, L.; McLaughlin, M.; Watts-Williams, S.J. The mycorrhizal pathway of zinc uptake contributes to zinc accumulation in barley and wheat grain. *BMC Plant. Biol.* **2019**, *19*, 133. [CrossRef] [PubMed]

41. Conversa, G.; Lazzizera, C.; Chiaravalle, A.E.; Miedico, O.; Bonasia, A.; La Rotonda, P.; Elia, A. Selenium fern application and arbuscular mycorrhizal fungi soil inoculation enhance Se content and antioxidant properties of green asparagus (Asparagus officinalis L.) spears. *Sci. Hortic.* **2019**, *252*, 176–191. [CrossRef]

42. Luo, W.; Li, J.; Ma, X.; Niu, H.; Hou, S.; Wu, F. Effect of arbuscular mycorrhizal fungi on uptake of selenate, selenite, and selenomethionine by roots of winter wheat. *Plant. Soil* **2019**, *438*, 71–83. [CrossRef]

43. Watts-Williams, S.J.; Gilbert, S.E. Arbuscular mycorrhizal fungi affect the concentration and distribution of nutrients in the grain differently in barley compared with wheat. *Plants People Planet* **2020**. [CrossRef]

44. Pellegrino, E.; Bedini, S. Enhancing ecosystem services in sustainable agriculture: Biofertilization and biofortification of chickpea (*Cicer arietinum* L.) by arbuscular mycorrhizal fungi. *Soil Biol. Biochem.* **2014**, *68*, 429–439. [CrossRef]

45. Lehmann, A.; Veresoglou, S.D.; Leifheit, E.F.; Rillig, M.C. Arbuscular mycorrhizal influence on zinc nutrition in crop plants—A meta-analysis. *Soil Biol. Biochem.* **2014**, *69*, 123–131. [CrossRef]

46. Burleigh, S.H.; Bechmann, I.E. Plant nutrient transporter regulation in arbuscular mycorrhizas. *Plant. Soil* **2002**, *244*, 247–251. [CrossRef]

47. Allen, J.W.; Shachar-Hill, Y. Sulfur Transfer through an Arbuscular Mycorrhiza. *Plant. Physiol.* **2008**, *149*, 549–560. [CrossRef]

48. Sieh, D.; Watanabe, M.; Devers, E.A.; Brueckner, F.; Hoefgen, R.; Krajinski, F. The arbuscular mycorrhizal symbiosis influences sulfur starvation responses of *Medicago truncatula*. *New Phytol.* **2012**, *197*, 606–616. [CrossRef]

49. Giovannetti, M.; Tolosano, M.; Volpe, V.; Kopriva, S.; Bonfante, P. Identification and functional characterization of a sulfate transporter induced by both sulfur starvation and mycorrhiza formation in *Lotus japonicus*. *New Phytol.* **2014**, *204*, 609–619. [CrossRef]

50. Nguyen, T.D.; Cavagnaro, T.R.; Watts-Williams, S.J. The effects of soil phosphorus and zinc availability on plant responses to mycorrhizal fungi: A physiological and molecular assessment. *Sci. Rep.* **2019**, *9*, 14880. [CrossRef]

51. Ma, J.; Janouskova, M.; Ye, L.; Bai, L.; Dong, R.; Yan, Y.; Yu, X.; Zou, Z.; Li, Y.; He, C. Role of arbuscular mycorrhiza in alleviating the effect of cold on the photosynthesis of cucumber seedlings. *Photosynthetica* **2019**, *57*, 86–95. [CrossRef]

52. Ma, X.; Luo, W.; Li, J.; Wu, F. Arbuscular mycorrhizal fungi increase both concentrations and bioavilability of Zn in wheat (*Triticum aestivum* L) grain on Zn-spiked soils. *Appl. Soil Ecol.* **2019**, *135*, 91–97. [CrossRef]

53. Gorzelak, M.A.; Asay, A.K.; Pickles, B.; Simard, S.W. Inter-Plant communication through mycorrhizal networks mediates complex adaptive behaviour in plant communities. *Aob Plants* **2015**, *7*, 7. [CrossRef]

54. Ingraffia, R.; Amato, G.; Frenda, A.S.; Giambalvo, D. Impacts of arbuscular mycorrhizal fungi on nutrient uptake, N2 fixation, N transfer, and growth in a wheat/faba bean intercropping system. *PLoS ONE* **2019**, *14*, e0213672. [CrossRef] [PubMed]

55. Lenoir, I.; Fontaine, J.; Sahraoui, A.L.-H. Arbuscular mycorrhizal fungal responses to abiotic stresses: A review. *Phytochemistry* **2016**, *123*, 4–15. [CrossRef] [PubMed]

56. Del Val, C.; Barea, J.M.; Azcón-Aguilar, C. Assessing the tolerance to heavy metals of arbuscular mycorrhizal fungi isolated from sewage sludge-contaminated soils. *Appl. Soil Ecol.* **1999**, *11*, 261–269. [CrossRef]

57. Del Val, C.; Barea, J.M.; Azcón-Aguilar, C. Diversity of Arbuscular Mycorrhizal Fungus Populations in Heavy-Metal-Contaminated Soils. *Appl. Environ. Microbiol.* **1999**, *65*, 718–723. [CrossRef]

58. Weissenhorn, I.; Glashoff, A.; Leyval, C.; Berthelin, J. Differential tolerance to Cd and Zn of arbuscular mycorrhizal (AM) fungal spores isolated from heavy metal-polluted and unpolluted soils. *Plant. Soil* **1994**, *167*, 189–196. [CrossRef]

59. Millar, N.S.; Bennett, A.E. Stressed out symbiotes: Hypotheses for the influence of abiotic stress on arbuscular mycorrhizal fungi. *Oecologia* **2016**, *182*, 625–641. [CrossRef]

60. Van Der Heijden, M.G.A.; Klironomos, J.N.; Ursic, M.; Moutoglis, P.; Streitwolf-Engel, R.; Boller, T.; Wiemken, A.; Sanders, I.R. Mycorrhizal fungal diversity determines plant biodiversity, ecosystem variability and productivity. *Nature* **1998**, *396*, 69–72. [CrossRef]

61. Bennett, A.E.; Classen, A. Climate change influences mycorrhizal fungal–plant interactions, but conclusions are limited by geographical study bias. *Ecology* **2020**, *101*. [CrossRef]

62. Kapoor, R.; Evelin, H.; Mathur, P.; Giri, B. Arbuscular Mycorrhiza: Approaches for Abiotic Stress Tolerance in Crop Plants for Sustainable Agriculture. In *Plant Acclimation to Environmental Stress*; Tuteja, N., Singh Gill, S., Eds.; Springer: New York, NY, USA, 2013; pp. 359–401. ISBN 978-1-4614-5001-6.

63. Porcel, R.; Aroca, R.; Ruiz-Lozano, J.M. Salinity stress alleviation using arbuscular mycorrhizal fungi. A review. *Agron. Sustain. Dev.* **2011**, *32*, 181–200. [CrossRef]

64. Hoeksema, J.D.; Chaudhary, V.B.; Gehring, C.A.; Johnson, N.C.; Karst, J.; Koide, R.T.; Pringle, A.; Zabinski, C.; Bever, J.D.; Moore, J.C.; et al. A meta-analysis of context-dependency in plant response to inoculation with mycorrhizal fungi. *Ecol. Lett.* **2010**, *13*, 394–407. [CrossRef]

65. Osakabe, Y.; Osakabe, K.; Shinozaki, K.; Tran, L.-S.P. Response of plants to water stress. *Front. Plant. Sci.* **2014**, *5*, 5. [CrossRef] [PubMed]

66. Balestrini, R.; Lumini, E. Focus on mycorrhizal symbioses. *Appl. Soil Ecol.* **2018**, *123*, 299–304. [CrossRef]

67. Bernardo, L.; Carletti, P.; Badeck, F.; Rizza, F.; Morcia, C.; Ghizzoni, R.; Rouphael, Y.; Colla, G.; Terzi, V.; Lucini, L. Metabolomic responses triggered by arbuscular mycorrhiza enhance tolerance to water stress in wheat cultivars. *Plant. Physiol. Biochem.* **2019**, *137*, 203–212. [CrossRef] [PubMed]

68. Augé, R.M. Water relations, drought and vesicular-arbuscular mycorrhizal symbiosis. *Mycorrhiza* **2001**, *11*, 3–42. [CrossRef]

69. Bárzana, G.; Aroca, R.; Paz, J.A.; Chaumont, F.; Martínez-Ballesta, M.C.; Carvajal, M.; Ruiz-Lozano, J.M. Arbuscular mycorrhizal symbiosis increases relative apoplastic water flow in roots of the host plant under both well-watered and drought stress conditions. *Ann. Bot.* **2012**, *109*, 1009–1017. [CrossRef]

70. Chitarra, W.; Maserti, B.; Gambino, G.; Guerrieri, E.; Balestrini, R. Arbuscular mycorrhizal symbiosis-mediated tomato tolerance to drought. *Plant. Signal. Behav.* **2016**, *11*, e1197468. [CrossRef]

71. Chitarra, W.; Pagliarani, C.; Maserti, B.; Lumini, E.; Siciliano, I.; Cascone, P.; Schubert, A.; Gambino, G.; Balestrini, R.; Guerrieri, E. Insights on the Impact of Arbuscular Mycorrhizal Symbiosis on Tomato Tolerance to Water Stress1. *Plant. Physiol.* **2016**, *171*, 1009–1023. [CrossRef]

72. Dell'Amico, J.; Torrecillas, A.; Rodríguez, P.; Morte, A.; Sánchez-Blanco, M.J. Responses of tomato plants associated with the arbuscular mycorrhizal fungus *Glomus clarum* during drought and recovery. *J. Agric. Sci.* **2002**, *138*, 387–393. [CrossRef]

73. Subramanian, K.; Santhanakrishnan, P.; Balasubramanian, P. Responses of field grown tomato plants to arbuscular mycorrhizal fungal colonization under varying intensities of drought stress. *Sci. Hortic.* **2006**, *107*, 245–253. [CrossRef]

74. Ruiz-Lozano, J.M.; Aroca, R.; Zamarreño, Á.M.; Molina, S.; Andreo-Jimenez, B.; Porcel, R.; García-Mina, J.M.; Ruyter-Spira, C.; López-Ráez, J.A. Arbuscular mycorrhizal symbiosis induces strigolactone biosynthesis under drought and improves drought tolerance in lettuce and tomato. *PlantCell Environ.* **2015**, *39*, 441–452. [CrossRef]

75. Allen, M.F.; Boosalis, M.G. Effects of two species of VA mycorrhizal fungi on drought tolerance of winter wheat. *New Phytol.* **1983**, *93*, 67–76. [CrossRef]

76. Marulanda, A.; Porcel, R.; Barea, J.M.; Azcón, R. Drought Tolerance and Antioxidant Activities in Lavender Plants Colonized by Native Drought-tolerant or Drought-sensitive Glomus Species. *Microb. Ecol.* **2007**, *54*, 543–552. [CrossRef] [PubMed]

77. Nelsen, C.E.; Safir, G.R. Increased drought tolerance of mycorrhizal onion plants caused by improved phosphorus nutrition. *Planta* **1982**, *154*, 407–413. [CrossRef] [PubMed]

78. Ortiz, N.; Armada, E.; Duque, E.; Roldan, A.; Azcón, R. Contribution of arbuscular mycorrhizal fungi and/or bacteria to enhancing plant drought tolerance under natural soil conditions: Effectiveness of autochthonous or allochthonous strains. *J. Plant. Physiol.* **2015**, *174*, 87–96. [CrossRef]

79. Abbaspour, H.; Saeidi-Sar, S.; Afshari, H.; Abdel-Wahhab, M.; Abdel-Wahhab, M.A. Tolerance of Mycorrhiza infected Pistachio (*Pistacia vera* L.) seedling to drought stress under glasshouse conditions. *J. Plant. Physiol.* **2012**, *169*, 704–709. [CrossRef]

80. Osonubi, O.; Mulongoy, K.; Awotoye, O.O.; Atayese, M.O.; Okali, D.U.U. Effects of ectomycorrhizal and vesicular-arbuscular mycorrhizal fungi on drought tolerance of four leguminous woody seedlings. *Plant. Soil* **1991**, *136*, 131–143. [CrossRef]

81. Ouledali, S.; Ennajeh, M.; Ferrandino, A.; Khemira, H.; Schubert, A.; Secchi, F. Influence of arbuscular mycorrhizal fungi inoculation on the control of stomata functioning by abscisic acid (ABA) in drought-stressed olive plants. *S. Afr. J. Bot.* **2019**, *121*, 152–158. [CrossRef]

82. De Ollas, C.; Dodd, I.C. Physiological impacts of ABA-JA interactions under water-limitation. *Plant. Mol. Biol.* **2016**, *91*, 641–650. [CrossRef]

83. Yosefi, M.; Sharafzadeh, S.; Bazrafshan, F.; Zare, M.; Amiri, B. Application of jasmonic acid can mitigate water deficit stress in cotton through yield-related physiological properties. *Acta Agrobot.* **2018**, *71*. [CrossRef]

84. Laxa, M.; Liebthal, M.; Telman, W.; Chibani, K.; Dietz, K.-J. The Role of the Plant Antioxidant System in Drought Tolerance. *Antioxidants* **2019**, *8*, 94. [CrossRef]

85. Mostofa, M.G.; Li, W.; Nguyen, K.H.; Fujita, M.; Tran, L.-S.P. Strigolactones in plant adaptation to abiotic stresses: An emerging avenue of plant research. *PlantCell Environ.* **2018**, *41*, 2227–2243. [CrossRef] [PubMed]

86. Padmavathi, T.; Dikshit, R.; Seshagiri, S. Influence of Rhizophagus spp. And Burkholderia seminalison the Growth of Tomato (*Lycopersicon esculatum*) and Bell Pepper (*Capsicum annuum*) under Drought Stress. *Commun. Soil Sci. Plant. Anal.* **2016**, *47*, 1975–1984. [CrossRef]

87. Khanam, D. Influence of Flooding on the Survival of Arbuscular Mycorrhiza. *Bangladesh J. Microbiol.* **2008**, *25*, 111–114. [CrossRef]

88. Wang, Y.; Huang, Y.; Qiu, Q.; Xin, G.; Yang, Z.; Shi, S. Flooding Greatly Affects the Diversity of Arbuscular Mycorrhizal Fungi Communities in the Roots of Wetland Plants. *PLoS ONE* **2011**, *6*, e24512. [CrossRef]

89. Bao, X.; Wang, Y.; Li, S.; Olsson, P.A. Arbuscular mycorrhiza under water—Carbon-phosphorus exchange between rice and arbuscular mycorrhizal fungi under different flooding regimes. *Soil Biol. Biochem.* **2019**, *129*, 169–177. [CrossRef]

90. Liang, W.; Ma, X.; Wan, P.; Liu, L. Plant salt-tolerance mechanism: A review. *Biochem. Biophys. Res. Commun.* **2018**, *495*, 286–291. [CrossRef]

91. Fougnies, L.; Renciot, S.; Müller, F.; Plenchette, C.; Prin, Y.; De Faria, S.M.; Bouvet, J.M.; Sylla, S.N.; Dreyfus, B.; Bâ, A.M. Arbuscular mycorrhizal colonization and nodulation improve flooding tolerance in *Pterocarpus officinalis* Jacq. seedlings. *Mycorrhiza* **2006**, *17*, 159–166. [CrossRef]

92. Wang, Y.; Qiu, Q.; Yang, Z.; Hu, Z.; Tam, N.; Xin, G. Arbuscular mycorrhizal fungi in two mangroves in South China. *Plant. Soil* **2009**, *331*, 181–191. [CrossRef]

93. Solís-Rodríguez, U.R.J.; Ramos-Zapata, J.; Hernández-Cuevas, L.; Salinas-Peba, L.; Guadarrama, P. Arbuscular mycorrhizal fungi diversity and distribution in tropical low flooding forest in Mexico. *Mycol. Prog.* **2020**, *19*, 195–204. [CrossRef]

94. Zhu, X.-C.; Song, F.-B.; Liu, S.-Q.; Liu, T.-D. Effects of arbuscular mycorrhizal fungus on photosynthesis and water status of maize under high temperature stress. *Plant. Soil* **2011**, *346*, 189–199. [CrossRef]

95. Caradonia, F.; Francia, E.; Morcia, C.; Ghizzoni, R.; Moulin, L.; Terzi, V.; Ronga, D. Arbuscular Mycorrhizal Fungi and Plant Growth Promoting Rhizobacteria Avoid Processing Tomato Leaf Damage during Chilling Stress. *Agronomy* **2019**, *9*, 299. [CrossRef]

96. Zhu, X.; Song, F.; Liu, F. Arbuscular Mycorrhizal Fungi and Tolerance of Temperature Stress in Plants. In *Arbuscular Mycorrhizas and Stress Tolerance of Plants*; Springer Science and Business Media LLC: Berlin, Germany, 2017; Volume 33, pp. 163–194.

97. Latef, A.A.H.A.; Chaoxing, H. Effect of arbuscular mycorrhizal fungi on growth, mineral nutrition, antioxidant enzymes activity and fruit yield of tomato grown under salinity stress. *Sci. Hortic.* **2011**, *127*, 228–233. [CrossRef]

98. Mathur, S.; Jajoo, A. Arbuscular mycorrhizal fungi protects maize plants from high temperature stress by regulating photosystem II heterogeneity. *Ind. Crop. Prod.* **2020**, *143*, 111934. [CrossRef]

99. Hajiboland, R.; Joudmand, A.; Aliasgharzad, N.; Tolrá, R.; Poschenrieder, C. Arbuscular mycorrhizal fungi alleviate low-temperature stress and increase freezing resistance as a substitute for acclimation treatment in barley. *Crop. Pasture Sci.* **2019**, *70*, 218–233. [CrossRef]

100. Bender, S.F.; Plantenga, F.; Neftel, A.; Jocher, M.; Oberholzer, H.-R.; Köhl, L.; Giles, M.; Daniell, T.J.; Van Der Heijden, M.G. Symbiotic relationships between soil fungi and plants reduce N_2O emissions from soil. *ISME J.* **2013**, *8*, 1336–1345. [CrossRef]

101. Berruti, A.; Lumini, E.; Balestrini, R.; Bianciotto, V. Arbuscular Mycorrhizal Fungi as Natural Biofertilizers: Let's Benefit from Past Successes. *Front. Microbiol.* **2016**, *6*. [CrossRef]

102. Gemma, J.N.; Koske, R.E. Seasonal Variation in Spore Abundance and Dormancy *of Gigaspora Gigantea* and in Mycorrhizal Inoculum Potential of a Dune Soil. *Mycology* **1988**, *80*, 211–216. [CrossRef]

103. Maya, M.A.; Matsubara, Y.-I. Influence of arbuscular mycorrhiza on the growth and antioxidative activity in cyclamen under heat stress. *Mycorrhiza* **2013**, *23*, 381–390. [CrossRef]

104. Chu, X.; Fu, J.; Sun, Y.; Xu, Y.; Miao, Y.; Xu, Y.; Hu, T. Effect of arbuscular mycorrhizal fungi inoculation on cold stress-induced oxidative damage in leaves of *Elymus nutans* Griseb. *S. Afr. J. Bot.* **2016**, *104*, 21–29. [CrossRef]

105. Yamato, M.; Ikeda, S.; Iwase, K. Community of arbuscular mycorrhizal fungi in a coastal vegetation on Okinawa island and effect of the isolated fungi on growth of sorghum under salt-treated conditions. *Mycorrhiza* **2008**, *18*, 241–249. [CrossRef]

106. Beltrano, J.; Ruscitti, M.; Arango, M.; Ronco, M. Effects of arbuscular mycorrhiza inoculation on plant growth, biological and physiological parameters and mineral nutrition in pepper grown under different salinity and p levels. *J. Soil Sci. Plant. Nutr.* **2013**, *13*, 123–141. [CrossRef]

107. Evelin, H.; Kapoor, R.; Giri, B. Arbuscular mycorrhizal fungi in alleviation of salt stress: A review. *Ann. Bot.* **2009**, *104*, 1263–1280. [CrossRef]

108. Amanifar, S.; Khodabandeloo, M.; Fard, E.M.; Askari, M.S.; Ashrafi, M. Alleviation of salt stress and changes in glycyrrhizin accumulation by arbuscular mycorrhiza in liquorice (*Glycyrrhiza glabra*) grown under salinity stress. *Environ. Exp. Bot.* **2019**, *160*, 25–34. [CrossRef]

109. Giri, B.; Kapoor, R.; Mukerji, K.G. Influence of arbuscular mycorrhizal fungi and salinity on growth, biomass, and mineral nutrition of *Acacia auriculiformis*. *Biol. Fertil. Soils* **2003**, *38*, 170–175. [CrossRef]

110. Djighaly, P.I.; Diagne, N.; Ngom, M.; Ngom, D.; Hocher, V.; Fall, D.; Diouf, D.; Laplaze, L.; Svistoonoff, S.; Champion, A. Selection of arbuscular mycorrhizal fungal strains to improve *Casuarina equisetifolia* L. and *Casuarina glauca* Sieb. tolerance to salinity. *Ann. Sci.* **2018**, *75*, 72. [CrossRef]

111. Hajiboland, R.; Aliasgharzad, N.; Laiegh, S.F.; Poschenrieder, C. Colonization with arbuscular mycorrhizal fungi improves salinity tolerance of tomato (*Solanum lycopersicum* L.) plants. *Plant. Soil* **2009**, *331*, 313–327. [CrossRef]

112. Daei, G.; Ardekani, M.; Rejali, F.; Teimuri, S.; Miransari, M. Alleviation of salinity stress on wheat yield, yield components, and nutrient uptake using arbuscular mycorrhizal fungi under field conditions. *J. Plant. Physiol.* **2009**, *166*, 617–625. [CrossRef]

113. Hashem, A.; Abd-Allah, E.F.; Alqarawi, A.A.; Al-Huqail, A.A.; Wirth, S.; Egamberdieva, D. The Interaction between Arbuscular Mycorrhizal Fungi and Endophytic Bacteria Enhances Plant Growth of *Acacia gerrardii* under Salt Stress. *Front. Microbiol.* **2016**, *7*. [CrossRef]

114. Li, J.; Meng, B.; Chai, H.; Yang, X.; Song, W.; Li, S.; Lu, A.; Zhang, T.; Sun, W. Arbuscular Mycorrhizal Fungi Alleviate Drought Stress in C3 (*Leymus chinensis*) and C4 (*Hemarthria altissima*) Grasses via Altering Antioxidant Enzyme Activities and Photosynthesis. *Front. Plant. Sci.* **2019**, *10*, 10. [CrossRef]

115. Wani, S.H.; Kumar, V.; Khare, T.; Guddimalli, R.; Parveda, M.; Solymosi, K.; Suprasanna, P.; Kishor, P.B.K. Engineering salinity tolerance in plants: Progress and prospects. *Planta* **2020**, *251*, 76. [CrossRef] [PubMed]

116. Talaat, N.B.; Shawky, B. Protective effects of arbuscular mycorrhizal fungi on wheat (*Triticum aestivum* L.) plants exposed to salinity. *Environ. Exp. Bot.* **2014**, *98*, 20–31. [CrossRef]

117. Munns, R.; Tester, M. Mechanisms of Salinity Tolerance. *Annu. Rev. Plant. Biol.* **2008**, *59*, 651–681. [CrossRef] [PubMed]

118. Li, Z.; Wu, N.; Meng, S.; Wu, F.; Liu, T. Arbuscular mycorrhizal fungi (AMF) enhance the tolerance of Euonymus maackii Rupr. at a moderate level of salinity. *PLoS ONE* **2020**, *15*, e0231497. [CrossRef]

119. Estrada, B.; Aroca, R.; Maathuis, F.J.M.; Barea, J.M.; Ruiz-Lozano, J.M. Arbuscular mycorrhizal fungi native from a Mediterranean saline area enhance maize tolerance to salinity through improved ion homeostasis. *PlantCell Environ.* **2013**, *36*, 1771–1782. [CrossRef]

120. Pedranzani, H.; Rodríguez-Rivera, M.; Gutierrez, M.; Porcel, R.; Hause, B.; Ruiz-Lozano, J.M. Arbuscular mycorrhizal symbiosis regulates physiology and performance of *Digitaria eriantha* plants subjected to abiotic stresses by modulating antioxidant and jasmonate levels. *Mycorrhiza* **2015**, *26*, 141–152. [CrossRef]

121. Al-Karaki, G.N. Nursery inoculation of tomato with arbuscular mycorrhizal fungi and subsequent performance under irrigation with saline water. *Sci. Hortic.* **2006**, *109*, 1–7. [CrossRef]

122. Giri, B.; Kapoor, R.; Mukerji, K.G. Improved Tolerance of *Acacia nilotica* to Salt Stress by Arbuscular Mycorrhiza, *Glomus fasciculatum* may be Partly Related to Elevated K/Na Ratios in Root and Shoot Tissues. *Microb. Ecol.* **2007**, *54*, 753–760. [CrossRef]

123. Soliman, A.S.; Shanan, N.T.; Massoud, O.N.; Swelim, D.M. Improving salinity tolerance of Acacia saligna (Labill.) plant by arbuscular mycorrhizal fungi and Rhizobium inoculation. *Afr. J. Biotechnol.* **2012**, *11*, 1259–1266. [CrossRef]

124. Santander, C.; Sanhueza, M.; Olave, J.; Borie, F.; Valentine, A.; Cornejo, P. Arbuscular Mycorrhizal Colonization Promotes the Tolerance to Salt Stress in Lettuce Plants through an Efficient Modification of Ionic Balance. *J. Soil Sci. Plant. Nutr.* **2019**, *19*, 321–331. [CrossRef]

125. Diouf, D.; Duponnois, R.; Ba, A.T.; Neyra, M.; Lesueur, D. Symbiosis of *Acacia auriculiformis* and *Acacia mangium* with mycorrhizal fungi and Bradyrhizobium spp. improves salt tolerance in greenhouse conditions. *Funct. Plant. Biol.* **2005**, *32*, 1143–1152. [CrossRef]

126. Wang, Y.-H.; Wang, M.; Li, Y.; Wu, A.; Huang, J. Effects of arbuscular mycorrhizal fungi on growth and nitrogen uptake of *Chrysanthemum morifolium* under salt stress. *PLoS ONE* **2018**, *13*, e0196408. [CrossRef]

127. Tian, C.; Feng, G.; Li, X.; Zhang, F. Different effects of arbuscular mycorrhizal fungal isolates from saline or non-saline soil on salinity tolerance of plants. *Appl. Soil Ecol.* **2004**, *26*, 143–148. [CrossRef]

128. Hashem, A.; Alqarawi, A.A.; Radhakrishnan, R.; Al-Arjani, A.-B.F.; Aldehaish, H.A.; Egamberdieva, D.; Allah, E.A. Arbuscular mycorrhizal fungi regulate the oxidative system, hormones and ionic equilibrium to trigger salt stress tolerance in *Cucumis sativus* L. *Saudi J. Biol. Sci.* **2018**, *25*, 1102–1114. [CrossRef] [PubMed]

129. Hashem, A.; Allah, E.A.; Alqarawi, A.A.; Wirth, S.; Egamberdieva, D. Comparing symbiotic performance and physiological responses of two soybean cultivars to arbuscular mycorrhizal fungi under salt stress. *Saudi J. Biol. Sci.* **2016**, *26*, 38–48. [CrossRef] [PubMed]

130. Hildebrandt, U.; Regvar, M.; Bothe, H. Arbuscular mycorrhiza and heavy metal tolerance. *Phytochemistry* **2007**, *68*, 139–146. [CrossRef]

131. Wang, F. Occurrence of arbuscular mycorrhizal fungi in mining-impacted sites and their contribution to ecological restoration: Mechanisms and applications. *Crit. Rev. Environ. Sci. Technol.* **2017**, *47*, 1–57. [CrossRef]

132. Kelkar, T.S.; Bhalerao, S.A. Beneficiary Effect of Arbuscular Mycorrhiza to *Trigonella Foenum-Graceum* in Contaminated Soil by Heavy Metal. *Res. J. Recent Sci.* **2013**, *2*, 29–32.

133. Kaldorf, M.; Kuhn, A.; Schröder, W.; Hildebrandt, U.; Bothe, H. Selective Element Deposits in Maize Colonized by a Heavy Metal Tolerance Conferring Arbuscular Mycorrhizal Fungus. *J. Plant. Physiol.* **1999**, *154*, 718–728. [CrossRef]

134. Gaur, A.; Adholeya, A. Prospects of arbuscular mycorrhizal fungi in phytoremediation of heavy metal contaminated soils. *Curr. Sci.* **2004**, *86*, 528–534.

135. Gong, X.; Tian, D.Q. Study on the effect mechanism of Arbuscular Mycorrhiza on the absorption of heavy metal elements in soil by plants. *IOP Conf. Ser. Earth Environ. Sci.* **2019**, *267*, 052064. [CrossRef]

136. Audet, P.; Charest, C. Heavy metal phytoremediation from a meta-analytical perspective. *Environ. Pollut.* **2007**, *147*, 231–237. [CrossRef] [PubMed]

137. Liu, L.; Li, J.; Yue, F.; Yan, X.; Wang, F.; Bloszies, S.; Wang, Y. Effects of arbuscular mycorrhizal inoculation and biochar amendment on maize growth, cadmium uptake and soil cadmium speciation in Cd-contaminated soil. *Chemosphere* **2018**, *194*, 495–503. [CrossRef] [PubMed]

138. González-Guerrero, M.; Escudero, V.; Saéz, Á.; Tejada-Jiménez, M. Transition Metal Transport in Plants and Associated Endosymbionts: Arbuscular Mycorrhizal Fungi and Rhizobia. *Front. Plant. Sci.* **2016**, *7*. [CrossRef] [PubMed]

139. Tamayo, E.; Ferrol, N.; Gómez-Gallego, T.; Azcón-Aguilar, C. Genome-wide analysis of copper, iron and zinc transporters in the arbuscular mycorrhizal fungus Rhizophagus irregularis. *Front. Plant. Sci.* **2014**, *5*, 5. [CrossRef]

140. Jiang, Q.-Y.; Zhuo, F.; Long, S.-H.; Zhao, H.-D.; Yang, D.-J.; Ye, Z.-H.; Li, S.-S.; Jing, Y.-X. Can arbuscular mycorrhizal fungi reduce Cd uptake and alleviate Cd toxicity of *Lonicera japonica* grown in Cd-added soils? *Sci. Rep.* **2016**, *6*, 21805. [CrossRef] [PubMed]

141. Hashem, A.; Allah, E.A.; Alqarawi, A.A.; Al Huqail, A.A.; Egamberdieva, D.; Wirth, S. Alleviation of cadmium stress in *Solanum lycopersicum* L. by arbuscular mycorrhizal fungi via induction of acquired systemic tolerance. *Saudi J. Biol. Sci.* **2016**, *23*, 272–281. [CrossRef] [PubMed]

142. Lingua, G.; Franchin, C.; Todeschini, V.; Castiglione, S.; Biondi, S.; Burlando, B.; Parravicini, V.; Torrigiani, P.; Berta, G. Arbuscular mycorrhizal fungi differentially affect the response to high zinc concentrations of two registered poplar clones. *Environ. Pollut.* **2008**, *153*, 137–147. [CrossRef]

143. Li, X.; Christie, P. Changes in soil solution Zn and pH and uptake of Zn by arbuscular mycorrhizal red clover in Zn-contaminated soil. *Chemosphere* **2001**, *42*, 201–207. [CrossRef]

144. Vos, C.; Tesfahun, A.; Panis, B.; De Waele, D.; Elsen, A. Arbuscular mycorrhizal fungi induce systemic resistance in tomato against the sedentary nematode *Meloidogyne incognita* and the migratory nematode *Pratylenchus penetrans*. *Appl. Soil Ecol.* **2012**, *61*, 1–6. [CrossRef]

145. Kumari, S.M.P.; Prabina, B.J. Protection of Tomato, *Lycopersicon esculentum* from Wilt Pathogen, *Fusarium oxysporum* f.sp. lycopersici by Arbuscular Mycorrhizal Fungi, Glomus sp. *Int. J. Curr. Microbiol. Appl. Sci.* **2019**, *8*, 1368–1378. [CrossRef]

146. Nguvo, K.J.; Gao, X. Weapons hidden underneath: Bio-Control agents and their potentials to activate plant induced systemic resistance in controlling crop Fusarium diseases. *J. Plant. Dis. Prot.* **2019**, *126*, 177–190. [CrossRef]

147. Spagnoletti, F.N.; Cornero, M.; Chiocchio, V.; Lavado, R.S.; Roberts, I.N. Arbuscular mycorrhiza protects soybean plants against *Macrophomina phaseolina* even under nitrogen fertilization. *Eur. J. Plant. Pathol.* **2020**, *156*, 839–849. [CrossRef]

148. Pozo, M.J.; Azcón-Aguilar, C. Unraveling mycorrhiza-induced resistance. *Curr. Opin. Plant. Biol.* **2007**, *10*, 393–398. [CrossRef]

149. Jung, R.E.; Zembic, A.; Pjetursson, B.E.; Zwahlen, M.; Thoma, D.S. Systematic review of the survival rate and the incidence of biological, technical, and aesthetic complications of single crowns on implants reported in longitudinal studies with a mean follow-up of 5 years. *Clin. Oral Implant. Res.* **2012**, *23*, 2–21. [CrossRef] [PubMed]

150. Cameron, D.D.; Neal, A.L.; Van Wees, S.A.; Ton, J. Mycorrhiza-induced resistance: More than the sum of its parts? *Trends Plant. Sci.* **2013**, *18*, 539–545. [CrossRef] [PubMed]

151. Siddiqui, Z.A.; Mahmood, I. Effect of a plant growth promoting bacterium, an AM fungus and soil types on the morphometrics and reproduction of *Meloidogyne javanica* on tomato. *Appl. Soil Ecol.* **1998**, *8*, 77–84. [CrossRef]

152. Akhtar, M.S.; Siddiqui, Z.A. Glomus intraradices, Pseudomonas alcaligenes, and Bacillus pumilus: Effective agents for the control of root-rot disease complex of chickpea (*Cicer arietinum* L.). *J. Gen. Plant. Pathol.* **2007**, *74*, 53–60. [CrossRef]

153. Hoffmann, D.; Vierheilig, H.; Schausberger, P. Arbuscular mycorrhiza enhances preference of ovipositing predatory mites for direct prey-related cues. *Physiol. Èntomol.* **2010**, *36*, 90–95. [CrossRef]

154. Vos, C.; Claerhout, S.; Mkandawire, R.; Panis, B.; De Waele, D.; Elsen, A. Arbuscular mycorrhizal fungi reduce root-knot nematode penetration through altered root exudation of their host. *Plant. Soil* **2011**, *354*, 335–345. [CrossRef]

155. Song, Y.; Chen, D.; Lu, K.; Sun, Z.; Zeng, R. Enhanced tomato disease resistance primed by arbuscular mycorrhizal fungus. *Front. Plant. Sci.* **2015**, *6*, 6. [CrossRef] [PubMed]

156. El-Khallal, S.M. Induction and modulation of resistance in tomato plants against Fusarium wilt disease by bioagent fungi (arbuscular mycorrhiza) and/or hormonal elicitors (Jasmonic acid & Salicylic acid): 2-Changes in the antioxidant enzymes, phenolic compounds and pathogen related- proteins. *Aust. J. Basic Appl. Sci.* **2007**, *1*, 717–732.

157. Akhtar, M.S.; Siddiqui, Z.A.; Wiemken, A. Arbuscular Mycorrhizal Fungi and Rhizobium to Control Plant Fungal Diseases. *Altern. Farming Syst. Biotechnol. Drought Stress Ecol. Fertil.* **2010**, 263–292. [CrossRef]

158. Gallou, A.; Mosquera, H.P.L.; Cranenbrouck, S.; Suárez, J.P.; Declerck, S. Mycorrhiza induced resistance in potato plantlets challenged by *Phytophthora infestans*. *Physiol. Mol. Plant. Pathol.* **2011**, *76*, 20–26. [CrossRef]

159. Koricheva, J.; Gange, A.C.; Jones, T. Effects of mycorrhizal fungi on insect herbivores: A meta-analysis. *Ecology* **2009**, *90*, 2088–2097. [CrossRef] [PubMed]

160. Gange, A.C.; Brown, V.K.; Aplin, D.M. Multitrophic links between arbuscular mycorrhizal fungi and insect parasitoids. *Ecol. Lett.* **2003**, *6*, 1051–1055. [CrossRef]

161. Gange, A.C.; West, H.M. Interactions between arbuscular mycorrhizal fungi and foliar-feeding insects in *Plantago lanceolata* L. *New Phytol.* **1994**, *128*, 79–87. [CrossRef]

162. Atera, E.A.; Itoh, K.; Azuma, T.; Ishii, T. Farmers' perspectives on the biotic constraint of *Striga hermonthica* and its control in western Kenya. *Weed Biol. Manag.* **2012**, *12*, 53–62. [CrossRef]

163. Jones, N.; Madhura, A.; Prashant, S.; Ramesh, B.; Jagadeesh, K.; Asha, A. Evaluation of arbuscular mycorrhizal fungi for suppression of *Striga hermonthica*, a parasitic weed in sorghum. In Proceedings of the Biennial Conference on Emerging Challenges in Weed Management, Jabalpur, India, 15–17 February 2014; p. 227.

164. Lendzemo, V.W.; Van Ast, A.; Kuyper, T.W. Can Arbuscular Mycorrhizal Fungi Contribute to Striga Management on Cereals in Africa? *Outlook Agric.* **2006**, *35*, 307–311. [CrossRef]

165. Manjunatha, H.P.; Jones Nirmalnath, P.; Chandranath, H.T.; Shiney, A.; Jagadeesh, K.S. Field evalualtion of native arbuscular mycorrhizal fungi in the management of Striga in sugarcane (*Saccharum officinarum* L.). *J. Pharm. Phytochem.* **2018**, *7*, 2496–2500.

166. Wang, Y.-Y.; Yin, Q.-S.; Qu, Y.; Li, G.-Z.; Hao, L. Arbuscular mycorrhiza-mediated resistance in tomato against *Cladosporium fulvum* -induced mould disease. *J. Phytopathol.* **2017**, *166*, 67–74. [CrossRef]

167. Liu, Y.; Feng, X.; Gao, P.; Li, Y.; Christensen, M.J.; Duan, T. Arbuscular mycorrhiza fungi increased the susceptibility of *Astragalus adsurgens* to powdery mildew caused by *Erysiphe pisi*. *Mycology* **2018**, *9*, 223–232. [CrossRef] [PubMed]

168. Kumari, S.M.P.; Srimeena, N. Arbuscular Mycorrhizal Fungi (AMF) Induced Defense Factors against the Damping-off Disease Pathogen, *Pythium aphanidermatum* in Chilli (*Capsicum annum*). *Int. J. Curr. Microbiol. Appl. Sci.* **2019**, *8*, 2243–2248. [CrossRef]

169. Singh, P.K.; Singh, M.; Vyas, D. Biocontrol of Fusarium Wilt of Chickpea using Arbuscular Mycorrhizal Fungi and *Rhizobium leguminosorum* Biovar. *Caryologia* **2010**, *63*, 349–353. [CrossRef]

170. Martínez-Medina, A.; Roldan, A.; Pascual, J.A. Interaction between arbuscular mycorrhizal fungi and *Trichoderma harzianum* under conventional and low input fertilization field condition in melon crops: Growth response and Fusarium wilt biocontrol. *Appl. Soil Ecol.* **2011**, *47*, 98–105. [CrossRef]

171. Kulkarni, S.A.; Kulkarni, S.; Sreenivas, M.N. Interaction Between Vesicular-Arbuscular (V-A) Mycorrhizae and *Sclerotium rolfsii* Sacc. in Groundnut. *J. Farm Sci.* **2011**, *10*, 919–921.

172. Jones, N.; Krishnaraj, P.; Kulkarni, J.; Patil, A.; Laxmipathy, R.; Vasudeva, R. Diversity of arbuscular mycorrhizal fungi in different ecological zones of northern Karnataka. *Eco Environ. Cons.* **2011**, *18*, 1053–1058.

173. Thiem, D.; Szmidt-Jaworska, A.; Baum, C.; Muders, K.; Niedojadło, K.; Hrynkiewicz, K. Interactive physiological response of potato (*Solanum tuberosum* L.) plants to fungal colonization and Potato virus Y (PVY) infection. *Acta Mycol.* **2014**, *1*, 291–303. [CrossRef]

174. Stolyarchuk, I.M.; Shevchenko, T.P.; Polischuk, V.P.; Kripka, A.V. Virus infection course in different plant species under influence of arbuscular mycorrhiza. *Microbiology* **2009**, *3*, 70–75. [CrossRef]

175. Othira, J.O. Effectiveness of arbuscular mycorrhizal fungi in protection of maize (*Zea mays* L.) against witchweed (*Striga hermonthica* Del Benth) infestation. *J. Agric. Biotechnol. Sustain. Dev.* **2012**, *4*, 37–44. [CrossRef]

176. Isah, K.; Kumar, N.; Lagoke, S.O.; Atayese, M. Management of *Striga hermonthica* on sorghum (*Sorghum bicolor*) using arbuscular mycorrhizal fungi (*Glomus mosae*) and NPK fertilizer levels. *Pak. J. Biol. Sci.* **2013**, *16*, 1563–1568. [CrossRef]

177. Wilson, G.; Rice, C.W.; Rillig, M.C.; Springer, A.; Hartnett, D.C. Soil aggregation and carbon sequestration are tightly correlated with the abundance of arbuscular mycorrhizal fungi: Results from long-term field experiments. *Ecol. Lett.* **2009**, *12*, 452–461. [CrossRef] [PubMed]

178. Giovannini, L.; Palla, M.; Agnolucci, M.; Avio, L.; Sbrana, C.; Turrini, A.; Giovannetti, M. Arbuscular Mycorrhizal Fungi and Associated Microbiota as Plant Biostimulants: Research Strategies for the Selection of the Best Performing Inocula. *Agronomy* **2020**, *10*, 106. [CrossRef]

179. Nacoon, S.; Jogloy, S.; Riddech, N.; Mongkolthanaruk, W.; Kuyper, T.W.; Boonlue, S. Interaction between Phosphate Solubilizing Bacteria and Arbuscular Mycorrhizal Fungi on Growth Promotion and Tuber Inulin Content of *Helianthus tuberosus* L. *Sci. Rep.* **2020**, *10*, 1–10. [CrossRef] [PubMed]

180. Vázquez, M.M.; César, S.; Azcón, R.; Barea, J.M. Interactions between arbuscular mycorrhizal fungi and other microbial inoculants (Azospirillum, Pseudomonas, Trichoderma) and their effects on microbial population and enzyme activities in the rhizosphere of maize plants. *Appl. Soil Ecol.* **2000**, *15*, 261–272. [CrossRef]

181. Khan, M.S.; Zaidi, A. Synergistic Effects of the Inoculation with Plant Growth-Promoting Rhizobacteria and an Arbuscular Mycorrhizal Fungus on the Performance of Wheat. *Turk. J. Agric.* **2007**, *31*, 355–362.

182. Diagne, N.; Baudoin, E.; Svistoonoff, S.; Ouattara, C.; Diouf, D.; Kane, A.; Ndiaye, C.; Noba, K.; Bogusz, D.; Franche, C.; et al. Effect of native and allochthonous arbuscular mycorrhizal fungi on Casuarina equisetifolia growth and its root bacterial community. *Arid. Land Res. Manag.* **2017**, *32*, 212–228. [CrossRef]

183. Nanjundappa, A.; Bagyaraj, D.J.; Saxena, A.K.; Kumar, M.; Chakdar, H. Interaction between arbuscular mycorrhizal fungi and Bacillus spp. in soil enhancing growth of crop plants. *Fungal Biol. Biotechnol.* **2019**, *6*, 1–10. [CrossRef]

184. Barea, J.M.; Azcón, R.; Azcón-Aguilar, C. Mycorrhizosphere interactions to improve plant fitness and soil quality. *Antonie Leeuwenhoek* **2002**, *81*, 343–351. [CrossRef]

185. Wu, Q.-S.; Tang, T.-T.; Xie, M.-M.; Chen, S.-M.; Zhang, S.-M. Effects of Arbuscular Mycorrhizal Fungi and Rhizobia on Physiological Activities in White Clover (*Trifolium repens*). *Biotechnology* **2019**, *18*, 49–54. [CrossRef]

186. Larimer, A.L.; Clay, K.; Bever, J.D. Synergism and context dependency of interactions between arbuscular mycorrhizal fungi and rhizobia with a prairie legume. *Ecology* **2014**, *95*, 1045–1054. [CrossRef]

187. Xavier, L. Response of lentil under controlled conditions to co-inoculation with arbuscular mycorrhizal fungi and rhizobia varying in efficacy. *Soil Biol. Biochem.* **2002**, *34*, 181–188. [CrossRef]

188. Wang, X.; Pan, Q.; Chen, F.; Yan, X.; Liao, H. Effects of co-inoculation with arbuscular mycorrhizal fungi and rhizobia on soybean growth as related to root architecture and availability of N and P. *Mycorrhiza* **2010**, *21*, 173–181. [CrossRef]

189. Bagyaraj, D.J.; Manjunath, A.; Patil, R.B. Interaction between a vesicular-arbuscular mycorrhiza and rhizobium and their effects on soybean in the field. *New Phytol.* **1979**, *82*, 141–145. [CrossRef]

190. Xavier, L.J.C.; Germida, J.J. Selective interactions between arbuscular mycorrhizal fungi and *Rhizobium leguminosarum* bv. viceae enhance pea yield and nutrition. *Biol. Fertil. Soils* **2003**, *37*, 261–267. [CrossRef]

191. Abd-Alla, M.H.; El-Enany, A.-W.E.; Nafady, N.A.; Khalaf, D.M.; Morsy, F.M. Synergistic interaction of *Rhizobium leguminosarum* bv. viciae and arbuscular mycorrhizal fungi as a plant growth promoting biofertilizers for faba bean (*Vicia faba* L.) in alkaline soil. *Microbiol. Res.* **2014**, *169*, 49–58. [CrossRef]

192. Jin, L.; Sun, X.; Wang, X.; Shen, Y.; Hou, F.; Chang, S.; Wang, C. Synergistic interactions of arbuscular mycorrhizal fungi and rhizobia promoted the growth of *Lathyrus sativus* under sulphate salt stress. *Symbiosis* **2010**, *50*, 157–164. [CrossRef]

193. Chatarpaul, L.; Chakravarty, P.; Subramaniam, P. Studies in tetrapartite symbioses. *Plant. Soil* **1989**, *118*, 145–150. [CrossRef]

194. Herrera, M.A.; Salamanca, C.P.; Barea, J.M. Inoculation of Woody Legumes with Selected Arbuscular Mycorrhizal Fungi and Rhizobia to Recover Desertified Mediterranean Ecosystems. *Appl. Environ. Microbiol.* **1993**, *59*, 129–133. [CrossRef]

195. Orfanoudakis, M.Z.; Hooker, J.E.; Wheeler-Jones, C.T. Early interactions between arbuscular mycorrhizal fungi and Frankia during colonisation and root nodulation of *Alnus glutinosa*. *Symbiosis* **2004**, *36*, 69–82.

196. Orfanoudakis, M.; Wheeler, C.T.; Hooker, J.E. Both the arbuscular mycorrhizal fungus *Gigaspora rosea* and Frankia increase root system branching and reduce root hair frequency in *Alnus glutinosa*. *Mycorrhiza* **2009**, *20*, 117–126. [CrossRef] [PubMed]

197. Oliveira, R.S.; Castro, P.M.L.; Dodd, J.; Vosátka, M. Synergistic effect of *Glomus intraradices* and Frankia spp. on the growth and stress recovery of *Alnus glutinosa* in an alkaline anthropogenic sediment. *Chemosphere* **2005**, *60*, 1462–1470. [CrossRef] [PubMed]

198. Andrade, D.D.S.; Leal, A.C.; Ramos, A.L.M.; De Goes, K.C.G.P. Growth of *Casuarina cunninghamiana* inoculated with arbuscular mycorrhizal fungi and Frankia actinomycetes. *Symbiosis* **2015**, *66*, 65–73. [CrossRef]

199. Chonglu, Z.; Mingqin, G.; Yu, C.; Fengzhen, W. Inoculation of Casuarina with Ectomycorrhizal Fungi, Vesicular-Arbuscular Mycorrhizal Fungi and Frankia. In Mycorrhizas for Plantation Forestry in Asia—ACIAR. In Proceedings of the International Symposium and Workshop, Kaiping, China, 7–11 November 2014; p. 122.

200. Wheeler, C.; Tilak, M.; Scrimgeour, C.; Hooker, J.; Handley, L. Effects of symbiosis with Frankia and arbuscular mycorrhizal fungus on the natural abundance of 15N in four species of Casuarina. *J. Exp. Bot.* **2000**, *51*, 287–297. [CrossRef] [PubMed]

201. Visser, S.; Danielson, R.M.; Parkinson, D. Field performance of *Elaeagnus commutata* and *Shepherdia canadensis* (*Elaeagnaceae*) inoculated with soil containing Frankia and vesicular–arbuscular mycorrhizal fungi. *Can. J. Bot.* **1991**, *69*, 1321–1328. [CrossRef]

202. Cely, M.V.T.; Siviero, M.A.; Emiliano, J.; Spago, F.R.; Freitas, V.F.; Barazetti, A.R.; Goya, E.T.; Lamberti, G.D.S.; Dos Santos, I.M.O.; De Oliveira, A.G.; et al. Inoculation of *Schizolobium parahyba* with Mycorrhizal Fungi and Plant Growth-Promoting Rhizobacteria Increases Wood Yield under Field Conditions. *Front. Plant. Sci.* **2016**, *7*. [CrossRef]

203. Vafadar, F.; Amooaghaie, R.; Otroshy, M. Effects of plant-growth-promoting rhizobacteria and arbuscular mycorrhizal fungus on plant growth, stevioside, NPK, and chlorophyll content of *Stevia rebaudiana*. *J. Plant. Interact.* **2013**, *9*, 128–136. [CrossRef]

204. Karthikeyan, B.; Abitha, B.; Henry, A.J.; Sa, T.; Joe, M.M. Interaction of Rhizobacteria with Arbuscular Mycorrhizal Fungi (AMF) and Their Role in Stress Abetment in Agriculture. In *Fungal Biology*; Springer Science and Business Media LLC: Berlin, Germany, 2016; pp. 117–142.
205. Aalipour, H.; Nikbakht, A.; Etemadi, N.; Rejali, F.; Soleimani, M. Biochemical response and interactions between arbuscular mycorrhizal fungi and plant growth promoting rhizobacteria during establishment and stimulating growth of Arizona cypress (*Cupressus arizonica* G.) under drought stress. *Sci. Hortic.* **2020**, *261*, 108923. [CrossRef]
206. Tank, N.; Saraf, M. Phosphate solubilization, exopolysaccharide production and indole acetic acid secretion by rhizobacteria isolated from *Trigonella foenum-graecum*. *Indian J. Microbiol.* **2003**, *43*, 37–40.
207. Hashem, A.; Abd_Allah, E.F.; Alqarawi, A.A.; Al-Huqail, A.A.; Shah, M.A. Induction of Osmoregulation and Modulation of Salt Stress in *Acacia gerrardii* Benth. by Arbuscular Mycorrhizal Fungi and *Bacillus subtilis* (BERA 71). *Biomed Res. Int.* **2016**, *2016*, 1–11. [CrossRef]
208. Lesueur, D.; Duponnois, R. Relations between rhizobial nodulation and root colonization of *Acacia crassicarpa* provenances by an arbuscular mycorrhizal fungus, *Glomus intraradices* Schenk and Smith or an ectomycorrhizal fungus, *Pisolithus tinctorius* Coker & Couch. *Ann. Sci.* **2005**, *62*, 467–474. [CrossRef]
209. Rajendran, K.; Devaraj, P. Biomass and nutrient distribution and their return of Casuarina equisetifolia inoculated with biofertilizers in farm land. *Biomass Bioenergy* **2004**, *26*, 235–249. [CrossRef]
210. Chilvers, G.A.; Lapeyrie, F.F.; Horan, D.P. Ectomycorrhizal vs Endomycorrhizal fungi within the same root system. *New Phytol.* **1987**, *107*, 441–448. [CrossRef]
211. Duponnois, R.; Diédhiou, S.; Chotte, J.L.; Sy, M.O. Relative importance of the endomycorrhizal and (or) ectomycorrhizal associations in Allocasuarina and Casuarina genera. *Can. J. Microbiol.* **2003**, *49*, 281–287. [CrossRef]
212. Tian, C.; He, X.; Zhong, Y.; Chen, J. Effect of inoculation with ecto- and arbuscular mycorrhizae and Rhizobium on the growth and nitrogen fixation by black locust, *Robinia pseudoacacia*. *New For.* **2003**, *25*, 125–131. [CrossRef]
213. Elumalai, S.; Raaman, N. In vitro synthesis of Frankia and mycorrhiza with *Casuarina equisetifolia* and ultrastructure of root system. *Indian J. Exp. Biol.* **2009**, *47*, 289–297.
214. Gunasekera, D.; Gunawardana, D.; Jayasinghearachchi, H. Diversity of Actinomycetes in Nitrogen Fixing Root Nodules of *Casuarina equisetifolia* and its Impact on Plant Growth. *Int. J. Multidiscip. Stud.* **2016**, *3*, 17. [CrossRef]
215. Duponnois, R. Mycorrhiza Helper Bacteria: Their Ecological Impact in Mycorrhizal Symbiosis. *Handb. Microb. Biofertil.* **2006**, *117*, 231–250.
216. Frey-Klett, P.; Garbaye, J.; Tarkka, M. The mycorrhiza helper bacteria revisited. *New Phytol.* **2007**, *176*, 22–36. [CrossRef]
217. Rigamonte, T.A.; Pylro, V.S.; Duarte, G.F. The role of mycorrhization helper bacteria in the establishment and action of ectomycorrhizae associations. *Braz. J. Microbiol.* **2010**, *41*, 832–840. [CrossRef]

Article

Definition of Core Bacterial Taxa in Different Root Compartments of *Dactylis glomerata*, Grown in Soil under Different Levels of Land Use Intensity

Jennifer Estendorfer [1], Barbara Stempfhuber [1,*], Gisle Vestergaard [1,2], Stefanie Schulz [1], Matthias C. Rillig [3], Jasmin Joshi [4], Peter Schröder [1] and Michael Schloter [1,5]

[1] Research Unit Comparative Microbiome Analysis, Helmholtz Zentrum München, 85764 Neuherberg, Germany
[2] Department of Health Technology, Technical University of Denmark, 2800 Lyngby, Denmark
[3] Institute for Biology, Freie Universität Berlin, 14195 Berlin, Germany
[4] Eastern Switzerland University of Applied Sciences, ILF Institute for Landscape and Open Space—Landscape Ecology, 8640 Rapperswil, Switzerland
[5] Chair for Soil Sciences, Technical University Munich, 85354 Freising, Germany
* Correspondence: Barbara.stempfhuber@helmholtz-muenchen.de; Tel.: +49-89-3187-2304

Received: 17 August 2020; Accepted: 2 October 2020; Published: 13 October 2020

Abstract: Plant-associated bacterial assemblages are critical for plant fitness. Thus, identifying a consistent plant-associated core microbiome is important for predicting community responses to environmental changes. Our target was to identify the core bacterial microbiome of orchard grass *Dactylis glomerata* L. and to assess the part that is most sensitive to land management. *Dactylis glomerata* L. samples were collected from grassland sites with contrasting land use intensities but comparable soil properties at three different timepoints. To assess the plant-associated bacterial community structure in the compartments rhizosphere, bulk soil and endosphere, a molecular barcoding approach based on high throughput 16S rRNA amplicon sequencing was used. A distinct composition of plant-associated core bacterial communities independent of land use intensity was identified. *Pseudomonas*, *Rhizobium* and *Bradyrhizobium* were ubiquitously found in the root bacterial core microbiome. In the rhizosphere, the majority of assigned genera were *Rhodoplanes*, *Methylibium*, *Kaistobacter* and *Bradyrhizobium*. Due to the frequent occurrence of plant-promoting abilities in the genera found in the plant-associated core bacterial communities, our study helps to identify "healthy" plant-associated bacterial core communities. The variable part of the plant-associated microbiome, represented by the fluctuation of taxa at the different sampling timepoints, was increased under low land use intensity. This higher compositional variation in samples from plots with low land use intensity indicates a more selective recruitment of bacteria with traits required at different timepoints of plant development compared to samples from plots with high land use intensity.

Keywords: land use intensity; plant-associated microbiome; endophytes; rhizosphere; biodiversity; bacteria; core microbiome; *Pseudomonas*

---

1. Introduction

It is generally accepted that microbiomes support plant growth and health at the plant soil interface [1–4]. The soil influenced by the root (rhizosphere) and the plant inner tissue (endosphere) provides distinct habitats for these microbial communities. In the rhizosphere, microorganisms benefit from exudation of organic compounds by the plant as well as mucilage provides ecological niches for important plant growth promoting microorganisms [5]. Moreover, plants secrete compounds to selectively chemoattract microorganisms, facilitating their colonization of and proliferation in the rhizosphere [6]. Endophytic bacteria are usually recruited from the rhizosphere microbiome and enter the plant either through lesions or penetrate the root surface actively in addition to vertical transmission of microbes via seeds [7,8]. Despite strong seasonal variations in relation to the plant development stage, the structure and function of plant-associated microbiomes are mostly influenced by the plant species [9].

However, in addition to the plant, microbial communities at the plant–soil interface are also shaped by other factors, including abiotic site-specific properties like soil characteristics, water availability and temperature. Furthermore, the type of management of a particular site and its intensity have been considered important drivers of the plant-associated microbial community [3,10–14]. We could recently show for *Dactylis glomerata* L., grown in a number of different grassland soils across Germany, that, independent of land use intensity (LUI), members of *Pseudomonadaceae*, *Enterobacteriaceae* and *Comamonadaceae* were the most abundant root endophytes, whereas in the rhizosphere and bulk soil, a clear influence of LUI on the microbial community structure was evident [15]. However, only a single date during peak vegetation was taken into account in this study, which might have masked effects of land use intensity on root endophytes due to the plants' impact as a result of high exudation rates at the selected sampling time.

Numerous studies postulate that despite the dynamic nature of the plant-associated microbial communities, plants may harbor a species-specific set of essential microbes that are not influenced by environmental conditions and plant growth, supporting the concept of plant-associated core microbial communities [16,17]. Those members may be crucial for nutrient uptake as well as the stress response of the plant and consequently determine the overall fitness of a plant [18]. Thus, the loss of parts of the plant-associated core microbial communities may induce reduced plant fitness and in the long run out-competition by other plants, which triggers significant shifts in biodiversity pattern worldwide, mainly if the plants are facing abiotic and biotic stressors. However, the consequences of LUI for the formation of a particular plant-associated core microbial composition at the root–soil interface are so far unclear.

In the frame of this study we selected eight grassland sites with different LUI within the Biosphere Reserve "Schwäbische Alb" in Southwestern Germany, to identify and define a putative root-associated bacterial core microbiome in (a) the endophytic compartment and (b) the rhizosphere of *D. glomerata*. Though plant-associated microbes comprise also of different entities such as fungi or protists, we focused in our study on plant-associated bacterial communities for which highly standardized analytical pipelines have been developed in the last years. For all sites, we analyzed bacterial diversity pattern using a molecular barcoding approach at different plant growth stages. We postulated that intensive land use (high LUI) may disentangle the close association of plants and microbes in terms of their co-occurrence and the composition of the core bacterial communities at the plant–soil interface (rhizosphere) might be less complex in terms of reduced bacterial diversity compared to sites with extensive forms of land use (low LUI). We expected that this effect be more pronounced in the rhizosphere than in the root interior.

2. Materials and Methods

2.1. Sampling Sites

The present study was conducted within the long-term interdisciplinary project of the German "Biodiversity Exploratories" (http://www.biodiversity-exploratories.de). Our study sites were located at the "Schwäbische Alb", which is a limestone secondary mountain range in the southwest of Germany, covering an area of about 422 km^2. The land use intensity of the experimental plots was assessed by a land use intensity index according to Blüthgen et al. [19], which was calculated for consecutive years (2006–2014) and included the three major management components fertilization, livestock density and mowing frequency (Table S1). The intensities of the different components were normalized to the regional mean. The index reflects a numerical gradient, formed by calculating the sum of the normalized LUI components equally weighted. For detailed equations please see Blüthgen et al. [19].

Plots were classified as high/low LUI based on their LUI index throughout several years (2006–2010). Selected sites were AEG6, AEG19, AEG20 and AEG21, representing high LUI plots and AEG7, AEG28, AEG33 and AEG34, representing low LUI plots (details can be seen in Table S1). Low LUI index ranged from 0.56 to 1.31 and high LUI index ranged from 1.55 to 2.25. The soil type of all plots was described as Rendzic Leptosol (according to the FAO classification system). Soil texture of the plots was determined as follows: clay content was in the range of 423–637 g/kg soil, silt content in the range of 327–554 g/kg soil and sand content in the range of 15–69 g/kg soil with exception of plot AEG7, showing differing values (clay: 385 g/kg soil, silt: 427 g/kg soil and sand: 188 g/kg soil). Total carbon contents of the sites were in a comparable range from 47 to 88 g/kg soil, total nitrogen contents were in the range of 4.8–10.6 g/kg soil. The C to N ratio was between 9 and 11 for all plots investigated. Samples were collected in May, June and October 2015, representing dynamic seasonal variations. The mean annual temperature was 9.9 °C and the annual precipitation was 730 mm. Mean temperature and precipitation during the months of sampling collection was 8.5–19 °C and 67.1 mm in May, 12.3–22.7 °C and 80.6 mm in June and 5.2–13.3 °C and 19.1 mm in October (Deutscher Wetterdienst, Offenbach, Station Stuttgart/Echterdingen). At each plot, samples were collected within a subplot of 1.5 m × 1.5 m. At each sampling timepoint, three plants of *D. glomerata* without any disease symptoms were excavated per plot and treated as true replicates.

2.2. Sampling and Basic Analyses

Roots of *D. glomerata* with adhering rhizosphere soil were suspended in 7.5 mL sterile 1× PBS solution amended with 0.02% Silwet (PBS, AppliChem, Darmstadt, Germany; Silwet L-77), and shaken at 180 rpm for 5 min to separate the rhizosphere from the roots. This step was repeated three times. The PBS solution containing the collected rhizosphere soil was centrifuged at 5000× *g* for 5 min and the pellet was frozen in liquid nitrogen and stored at −80 °C until further analyses.

After the separation of rhizosphere soil, roots were immediately surface sterilized. Therefore, roots were subjected to sterile 1% Tween 20 for 2 min and washed with pure autoclaved water. Next, roots were incubated for 2 min in 70% ethanol and rinsed three times in sterile distilled water. Subsequently, surface sterilization was done by incubating the roots in 5% sodium hypochlorite for 10 min and rinsing them in sterile water eight times. Sterilized roots were frozen in liquid nitrogen and stored at −80 °C. To check successful surface sterilization, DNA extraction followed by PCR amplification of the 16S rRNA genes, and incubation of 200 µL of the final rinse water on NB agar plates was performed. The absence of the PCR product and no colonies on the agar after 10 days at 28 °C confirmed successful sterilization.

Furthermore, bulk soil samples were taken (soil only loosely attached to the roots) and stored at −80 °C until further molecular analyses, respectively, sieved and stored for a maximum of 24 h at 4 °C for chemical analysis. Water extractable organic carbon (WEOC) and nitrogen (WEON) were determined using DIMA-TOC 100 (Dima Tec, Langenhagen, Germany) using 0.01 M calcium chloride solution for extraction [20]. The same extracts were used to measure nitrate (NO_3^--N) and ammonium (NH_4^+-N) by continuous flow analysis using a photometric autoanalyzer (CFA-SAN Plus; Skalar Analytik, Erkelenz, Germany).

In addition, above ground plant material was collected and rinsed with tap water, dried on 65 °C for 2 days and pulverized using the Tissue LyserII (Qiagen GmbH, Hilden, Germany). Total carbon and nitrogen contents were measured using an Elemental Analyzer 'Euro-EA' (Eurovector, Milano, Italy).

2.3. Nucleic Acid Extraction

For nucleic acid extraction, a phenol-chloroform-based method was used with slight modifications [21]. We used surface sterilized roots, the rhizosphere and bulk soil from all eight plots (3 replicates per plot) in May, June and October resulting in 216 samples in total. Prior to DNA extraction, surface sterilized roots were frozen in liquid nitrogen and prehomogenized using a TissueLyserII (Qiagen GmbH, Hilden, Germany). Afterwards, 0.1 g of roots and 0.3 g of rhizosphere and bulk soil were homogenized using lysing matrix tubes E (MP Biomedicals, Illkirch-Graffenstaden, France) in 120 mM sodium phosphate buffer (pH 8) and TNS solution (500 mM Tris-HCl pH 8.0, 100 mM NaCl, 10% SDS (wt/vol)), and centrifuged at 16,100× g for 10 min at 4 °C. The supernatant was transferred into a 2 mL DNase/RNase free SafeLock tube on ice and successively mixed with an equal volume of phenol/chloroform/isoamylalcohol (25:24:1 (vol/vol), Sigma-Aldrich, St. Louis, MO, USA) and chloroform/isoamylalcohol (24:1 (vol/vol)) and centrifuged for 5 min at 16,100× g. DNA was precipitated using 30% (wt/vol) polyethylene glycol (PEG) solution (PEG 6000, NaCl). After 2 h of incubation on ice, the solution was centrifuged (16,100× g, 10 min, 4 °C). The resulting pellet was washed in ice-cold DNase/RNase free 70% ethanol, air-dried and eluted in 30 μL 0.1% diethylpyrocarbonate water. The concentration was measured in duplicates using the Quant-iT™Pico Green® ds DNA assay Kit (Invitrogen, Carlsbad, CA, USA) according to the manufacturer's instructions. The measurements were performed with a SpectraMax Gemini EM Fluorescence Plate Reader Spectrometer (Molecular Devices, Sunnyvale, CA, USA). Values were corrected for background fluorescence by addition of negative controls. Finally, the DNA extracts were stored at −80 °C until further use.

2.4. Library Preparation and Illumina Sequencing

Next generation sequencing was performed on the Illumina MiSeq platform (Illumina Inc., San Diego, CA, USA). Library was prepared according to the "16S Metagenomic Sequencing Library Preparation" protocol proposed by Illumina Inc., USA. To reduce biases of the polymerase chain reaction (PCR) amplification of the 16S rRNA region, the reaction was carried out in triplicates using 335Fc (5′-CADACTCCTACGGGAGGC-3′) as a forward primer and 769Rc (5′-ATCCTGTTTGMTMCCCVCRC-3′) as a reverse primer with Illumina adapter sequences [22]. The PCR reaction contained 12.5 μL NEB Next High Fidelity Master Mix (Illumina Inc., San Diego, CA, USA), 0.5 μL of each primer (10 pmol/μL), 2.5 μL of 3% BSA, 100–200 ng of template DNA and 25 μL of DEPC water. PCR conditions included an initial denaturation step at 98 °C for 5 min, followed by 20 cycles (rhizosphere and bulk soil samples) or 28 cycles (root samples) of denaturation (98 °C; 10 s), annealing (60 °C; 30 s) and elongation (72 °C; 30 s). The final elongation was performed at 72 °C for 5 min. Negative controls of extraction (blank extraction) and PCR (using DEPC water instead of template DNA) were treated accordingly. Resulting amplicons were analyzed on a 2% agarose gel. Afterwards, triplicates were pooled and purified using the Agencourt®AMPure®XP (Beckman Coulter Company, Carlsbad, CA USA) extraction kit according to the manufacturer's instructions, with a modified ratio of AMPure XP to PCR reaction (0.6/1). The presence of primer-dimers and amplicon sizes were checked on a Bioanalyzer 2100 Agilent Technologies, Santa Clara, CA, USA), using the DNA 7500 kit (Agilent Technologies, Santa Clara, CA, USA) and quantified using the Quant-iT PicoGreen kit (Life Technologies, Grand Island, NY, USA). Finally, indexing PCR was carried out using 10 ng of amplicon DNA, 12.5 μL NEB Next High Fidelity Master Mix and 10 pmol of each indexing-primer. PCR conditions were changed for the annealing temperature (55 °C) and the number of cycles (8 cycles). Purified PCR products were pooled in equimolar ratios to a final concentration of 4 nM and sequenced using the MiSeq Reagent kit v3 (600 cycles; Illumina Inc., San Diego, CA, USA) for paired end sequencing. Sequence files were deposited in the NCBI Sequence Read Archive under accession numbers SRP102620 and PRJNA380810.

2.5. Sequence Data Analysis

Sequence analysis was performed using QIIME (quantitative insights into microbial ecology [23]) and default parameters. FASTQ files were trimmed and merged with a minimum read length of 50 and minimum Phred score of 15 using AdapterRemoval [24]. PhiX contamination was removed using DeconSeq [25]. Reads were merged and filtered by size (400–480 bp) and clustered into operational taxonomic units (OTUs) at 97% sequence identity with an open reference strategy using GreenGenes 16S rRNA reference database (13_5 release) [26]. Taxonomy was assigned using the RDP (v2.2) classifier retrained on the GreenGenes [27]. Afterwards, chloroplast sequences were removed, and output was filtered with an abundance cut-off of 0.001%. To make results comparable, the data set was rarefied to the lowest obtained read number. Afterwards, diversity analyses were performed, which is implemented in the Qiime workpackages, including the calculation of relative abundance of each OTU per sample as well as the computation of α- and ß-diversity. The analysis of α-diversity (within sample diversity) was calculated per sample and was based on chao1 richness [28] and Shannon's diversity [29]. For the calculation of the boxplots, samples were grouped by month and LUI (i.e., 3×4 samples high and 3×4 samples low LUI per sampling season and compartment). Significant differences in α-diversity were obtained by unpaired *t*-tests. Betadiversity measures were calculated using unweighted and weighted UniFrac metrics as described by Catherine et al. [30].

Statistical significance of dissimilarities in ß-diversity was determined by adonis using the r-package, Vegan (R package version 2.4-4) [31], via the Qiime script "compare_categories.py". Significances of LUI were calculated per compartment and season. Significances of season were calculated per compartment and LUI. The analysis of the core bacterial microbiome of roots and rhizosphere from different LUIs was based on the total relative abundance of bacterial OTUs and was computed using "compute_core_microbiome.py". Thus, a table of OTUs was obtained per month and LUI, where the OTUs that remained were present in 90% of the respective samples (i.e., OTU had to be present in 11 of 12 samples per LUI and month). These tables were used for visualization in the Bioinformatics and Evolutionary Genomics webtool [32].

3. Results

3.1. Soil Carbon and Nitrogen Content

While WEOC was not influenced by different LUI levels, it changed in response to the sampling season. Concentrations in October (64.1 μg g^{-1} dw in average) were significantly higher compared to May (35 μg g^{-1} dw in average) or June (26.1 μg g^{-1} dw in average). In contrast, WEON, nitrate and ammonium concentrations increased with LUI and changed over time. As expected, the highest concentrations of 43.13 μg g^{-1} dw WEON and 46.08 μg N g^{-1} dw nitrate on average were detected in June on intensively managed sites due to increased fertilizer input by manure application (see Table S2).

3.2. Sequencing Summary

In total, 16,085,722 raw-sequence reads were obtained from PCR amplicons by Illumina sequencing. After quality filtering and chimera check, 12,561,385 high-quality partial 16S rRNA gene sequences with a minimum length of 400 bp remained. After removal of chloroplasts and the application of an abundance cut-off of 0.001%, 8,991,033 sequences and 10,099 OTUs remained. To compare samples without statistical bias, data were rarefied to 14,092 reads per sample, which reflected the lowest obtained read number. Rarefaction was performed to account for variations in library sizes between samples of different compartments. Rarefaction analysis indicated a sufficient sampling depth for further investigation at 97% sequence similarity (Figure S1).

3.3. Characterization of Bacterial Diversity

Analysis of α-diversity (diversity in terms of OTU numbers present in a single sample), measured as the Shannon index and chao1 richness (Figure 1), indicated a significant impact of LUI mainly on bacterial diversity in bulk soil in June, with increased values at sites with high LUI compared to sites with low LUI. For other timepoints no significant influence of LUI on bacterial diversity was detected in bulk soil. For the other compartments (rhizosphere and root interior) no significant influence of LUI on bacterial diversity was measured for any of the sampling timepoints. However, as expected, overall α-diversity was higher in the rhizosphere and bulk soil compared to the root interior.

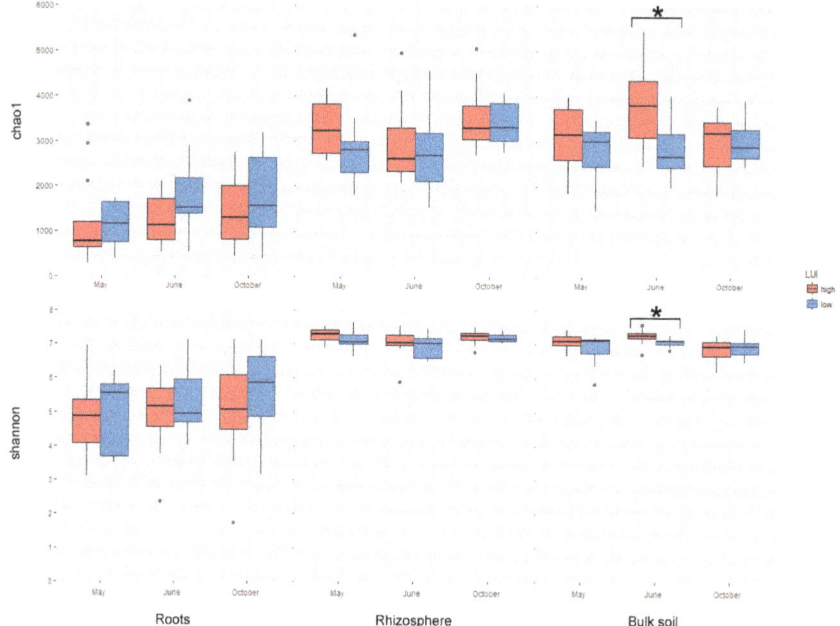

Figure 1. α-diversity measures chao1 species richness and Shannon index of all compartments under low and high land use intensity, respectively. The distribution is shown at 97% sequence similarity. The boxplot indicates the first and third quartile. The median is indicated as a horizontal line and whiskers indicate minimum or maximum, respectively. Significant differences among land use intensities are indicated with an asterisk, circles reflect outlier values (*t*-test, $p < 0.05$).

Beta-diversity was also impacted by LUI to a large extent, mainly in June, when effects of LUI on all compartments were observed. In addition, significant differences between different LUI levels were also observed in the rhizosphere and bulk soil for the other two sampling periods (Table S3). Weighted UniFrac distances indicated a significant impact of season for all analyzed compartments (Table S4).

Most abundant phyla in all compartments (endosphere, rhizosphere and bulk soil) were Proteobacteria, Bacteroidetes and Actinobacteria, followed by Firmicutes and Acidobacteria (Table S5). At the genus level (Table S6), *Pseudomonas* (23%) was dominating the root endosphere followed by *Janthinobacterium* (4%), *Rhizobium* (3%), *Burkholderia* (3%) and genera belonging to the family of Enterobacteriaceae (4%). In contrast, in the rhizosphere as well as in bulk soil, most abundant taxa were assigned to the families Sinobacteraceae (9/6%) and Chitinophagaceae (8/12%) followed by the genus *Rhodoplanes* (4/6%).

3.4. Definition of an Endophytic Core Microbiome for Dactylis glomerata L.

From the 88 OTUs detected in roots under high LUI on the 97% homology level, 76 were affected by the sampling timepoint. At all sampling timepoints, 12 OTUs were detected and formed a plant-associated bacterial core community (high LUI, May, June and October; Figure 2).

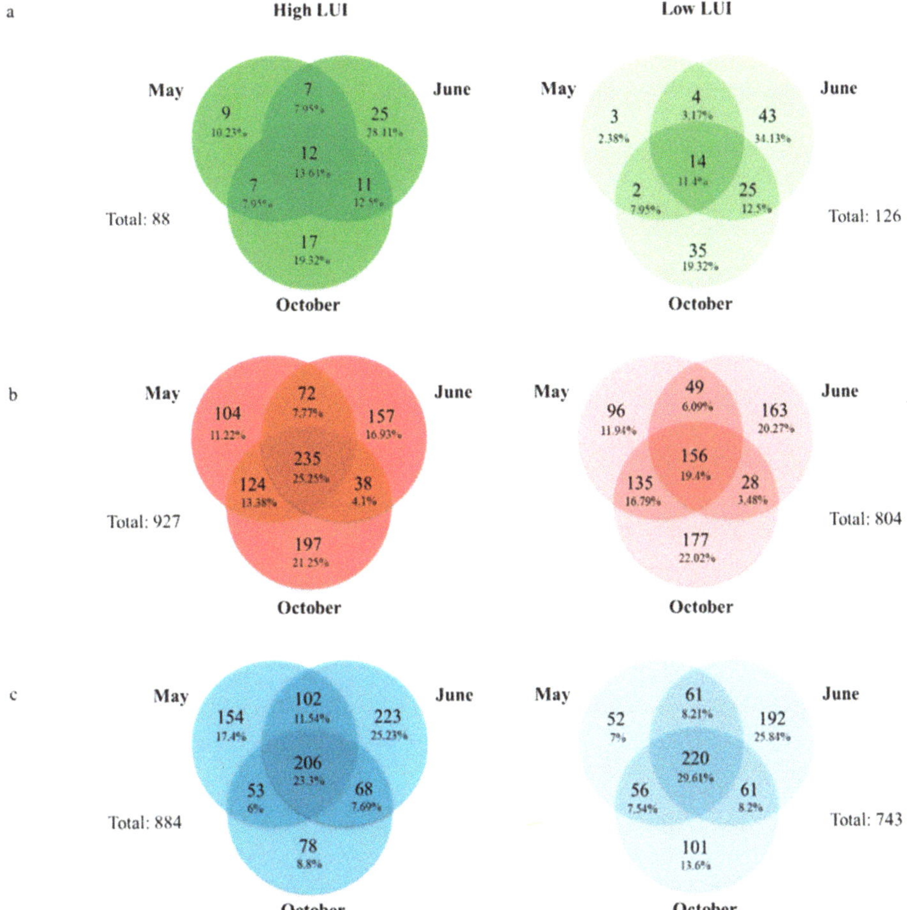

Figure 2. Shared operational taxonomic units (OTUs) between the different sampling seasons May, June and October for both land use intensities (LUIs). Green depicts root samples (**a**), red depicts rhizosphere samples (**b**) and blue depicts bulk soil samples (**c**). The percentages are based on the total amount of OTUs that contribute to the diagram.

Five of these OTUs were classified as *Pseudomonas* without a clear affiliation to a particular species (OTU 398604, 4455861, 3314521, 9448 and 1087); one OTU (633252) could be further classified as *P. veronii*. The other OTUs of the core microbiome could be assigned to *Rhizobium* (OTU 1104627, 220539), *Bradyrhizobium* (OTU 4377104, 1105814), *Agrobacterium* (OTU 969805) and *Labrys* (OTU 218772; see Table S7).

Overall, more OTUs could be detected in roots under low LUI compared to high LUI (126 OTUs compared to 88 on the level of 97% homology). However, the plant-associated core bacterial communities including all sampling timepoints (low LUI, May, June and October) were comparable (14 OTUs compared to 12; Figure 2). The classification revealed close similarities between the plant-associated core bacterial communities of high and low LUI. Like for high LUI, most OTUs in the core (8) could be identified as *Pseudomonas*. The OTU assigned to *P. veronii* for both land use intensities were identical (OTU 633252). Other identical OTUs found were OTUs 3,314,521, 9448 and 1087, classified also as *Pseudomonas* and OTU 4,377,104 and 1,105,814, assigned to *Bradyrhizobium*. OTUs that were unique in the core bacterial composition of plants from low LUI sites, but phylogenetically comparable to OTUs found in the roots of plants from high LUI sites were OTUs 12,056, 6025 and 14298, which were further classified to *P. umsongensis*. In addition to those, we found one OTU (646,549), classified as *Pseudomonas* and three OTUs classified as *Rhizobium* (2156, 1433, 1,104,627), which were phylogenetically similar to those detected in roots of plants from sites with high LUI.

The only OTU that was solely found as part of the root core microbiome of plants at sites with low LUIs was assigned to *Caulobacter* (OTU 7929). In contrast, the OTUs that were only present in the endophytic core microbiome of plants under high LUI were assigned to *Agrobacterium* (OTU 969805) and *Labrys* (OTU 218772; see Table S8).

3.5. Definition of a Core Microbiome for the Rhizosphere of Dactylis glomerata L.

Due to the higher diversity in the rhizosphere compared to the root interior, the absolute numbers of OTUs, which contributed to the bacterial core microbiome of the rhizosphere, were higher. However, surprisingly also relative numbers were increased (Figures 2b and 3b). While in the root interior the bacterial core composition was only formed by 11–13% of the detected OTUs, more than 19% of the detected OTUs were part of the plant-associated core bacterial communities in the rhizosphere (Figure 2a,b). Interestingly, the numbers, as well as the proportion of OTUs that contributed to the core that was shared amongst all sampling seasons in the rhizosphere was lower under low LUI (156/19.4%) compared to high LUI (235/25.3%). In contrast to the root core bacterial community composition (shared throughout all sampling seasons under high vs. low LUI), the amount of uniquely found OTUs was higher under high LUI compared to low LUI (Figure 3b).

While the core plant-associated bacteria (found across all sampling seasons) in roots were dominated by OTUs, which could be assigned to *Pseudomonas* under both LUIs, OTUs linked to this genus were less abundant in the core bacterial community (found across all sampling seasons) of the plant rhizosphere independent of LUI. Under both LUIs, the majority of core OTUs was assigned to the genera *Rhodoplanes*, *Methylibium*, *Kaistobacter* and *Bradyrhizobium*, and to the families Sinobacteraceae and Chitionophagaceae.

We found a total number of 927 OTUs under high LUI in the rhizosphere, where 692 OTUs were influenced by a sampling timepoint on a 97% sequence similarity level (not found across every season). 235 OTUs were detected independent of the sampling timepoint, reflecting the high LUI rhizosphere core (found across all seasons). The majority of OTUs found in the core was assigned to the genera *Rhodoplanes* (16 OTUs), *Methylibium* (15 OTUs), *Kaistobacter* (11 OTUs) and *Bradyrhizobium* (8 OTUs). Furthermore, a high number of not further assigned Sinobacteraceae (20 OTUs) and Chitinophagaceae (12 OTUs) was detected (Table S9).

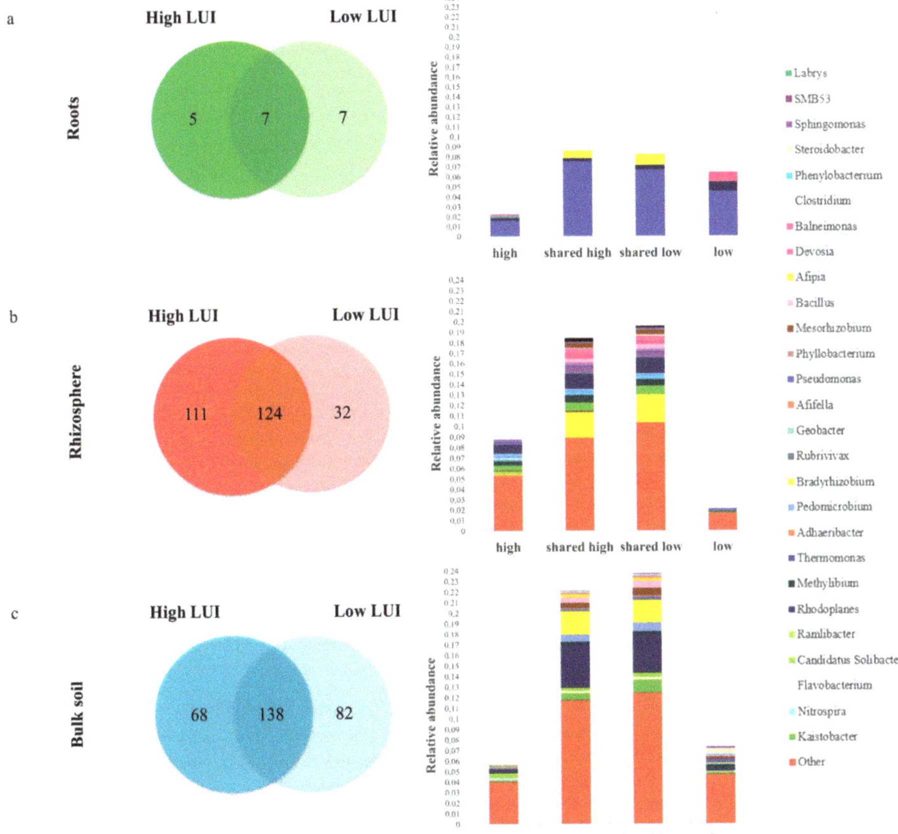

Figure 3. Comparison of plant-associated core bacterial communities. The Venn diagrams show the amount of OTUs that contributed to the core bacterial communities that were shared amongst all seasons only under high and low LUI as well as the core OTUs found under both LUIs and all seasons for roots (green, (**a**)), rhizosphere (red, (**b**)) and bulk soil (blue, (**c**)). The stackplots show the comparison of the total relative abundance of the OTUs that contribute to the core bacterial communities in Figure 3a. "High" refers to the abundance of OTUs found under high LUI only (shared amongst all seasons, but only under high LUI), "shared high" refers to the abundance under high LUI of those OTUs that were found under both LUIs (shared amongst all seasons under high as well as low LUI), "shared low" refers to the abundance under low LUI of those OTUs that were found under both LUIs (shared amongst all seasons under high as well as low LUI), abundance of OTUs under low LUI only (shared amongst all seasons, but only under low LUI), colored by genus. OTUs in "Other" could not be classified to the genus level.

Under low LUI, we found a total number of 804 OTUs, with 648 OTUs being influenced by the sampling date on the 97% sequence similarity level (not found across every season). The plant-associated core bacterial community composition was formed by 156 OTUs (found across all seasons). As for the root core plant-associated bacteria, the largest part of the rhizosphere core (low LUI) was comparable to the core found under high LUI regarding the genera detected. In total, we found 124 identical OTUs in both high and low LUI core bacterial community composition. The genera that were found in highest amounts were the same as under high LUI: *Rhodoplanes* (11 out of 12 OTUs identical to high LUI core), *Kaistobacter* (7 out of 8 OTUs identical), *Methylibium* (7 identical OTUs) and

Bradyrhizobium (7 identical OTUs). Additionally, not further classified families that were found in high numbers were the same: Sinobacteraceae (11 identical OTUs) and Chitinophagaceae (5 out of 8 OTUs identical). Thirty-two OTUs appeared to be unique in the rhizosphere core of low LUI. Among those were the genera *Pseudomonas* (OTUs 646,549, 398,604), *Candidatus Solibacter* (OTU 817,874), *Flavobacterium* (OTU 6614), *Kaistobacter* (OTU 331,282), *Niabella* (7994), *Novosphingobium* (OTU 941,803), *Pedobacter* (OTU 976,441) and *Rhodoplanes* (OTU 2,025,156). Moreover, six unique OTUs could be classified to a particular species (on the 97% homology level; Table S10, Table S11 and Table S12): *Asticcacaulis biprosthecium* (OTU 1,105,085), *Bosea genosp.* OTUs (567,840, 829,415), *Variovorax paradoxus* (OTUs 123, 2575) and *Sphingomonas wittichii* (OTU 3395). Additionally, we found three taxa that were only present in the low LUI plant-associated core bacterial community composition: *Novosphingobium* (OTU 94,180), *Niabella* (OTU 7994) and *Pedobacter* (OTU 976,441).

OTUs unique to the high LUI rhizosphere core (111) appeared to be more diverse compared to low LUI (Figure 3b). Within the 111 unique OTUs in the core of high LUI, 39 OTUs could be assigned to genus level without further classification to the species. Among those we found *Methylibium* (8 OTUs), *Rhodoplanes* (5 OTUs), *Kaistobacter* (4 OTUs) and *Candidatus Solibacter* (3 OTUs). Furthermore, among the unique OTUs under high LUI, we found OTUs that could be assigned to *Rhizobium* (103,410), *Mycoplana* (998,905), *Labrys* (543,156), *Asticcacaulis* (1,105,085) *Adhaeribacter* (1,069,076), *Paucibacter* (593,163) *Microlunatus* (249,330, 4209) and *Bosea* (567,840, 829,415).

3.6. Definition of a Core Microbiome for the Bulk Soil of Dactylis glomerata L.

The bulk soil associated bacterial core communities were comparable to the rhizosphere associated core bacterial communities with regard to the composition as well as to the total amount of shared OTUs found amongst all sampling seasons. Under high LUI, we found in total 884 OTUs, 678 of which were influenced by the sampling timepoint (not found across all seasons). We found 206 OTUs to be shared across all sampling seasons under high LUI on the 97% sequence similarity level. The majority of OTUs assigned to the genus level was also comparable to the core bacterial community composition of the rhizosphere. OTUs in the shared core of the bulk soil (found across all seasons) were assigned to *Rhodoplanes* (23 OTUs), *Kaistobacter* (10 OTUs), *Candidatus Solibacter* (8 OTUs), *Bradyrhizobium* (8 OTUs), *Methylibium* (8 OTUs) and *Pedomicrobium* (8 OTUs). High numbers of not further classified families were Sinobacteraceae (22 OTUs) and Chitinophagaceae (16 OTUs) like in the rhizosphere. Genera found to be unique in the core of high LUI were *Geobacter* (701,911), *Ramlibacter* (4928), Afifella (10,984), *Nitrospira* (173,004), *Adhaeribacter* (1,069,076) and *Thermomonas* (805,685). Only the latter two OTUs were also found in the high LUI core bacterial community composition of the rhizosphere.

Under low LUI, a total of 743 OTUs was detected across all seasons in the bulk soil. Among those, 220 could be found independent of the sampling timepoint (shared across all sampling seasons). Most OTUs could be classified to *Bacillus* (with 3 out of 5 OTUs being identical to high LUI OTUs), *Bradyrhizobium* (8 out of 9 identical OTUs), *Kaistobacter* (6 out of 8 identical OTUs), *Methylibium* (4 out of 9 identical OTUs), *Pedomicrobium* (7 out of 8 identical OTUs), *Rhodoplanes* (19 out of 22 identical OTUs), *Steroidobacter* (1 out of 7 identical OTUs), Sinobacteraceae (16 out of 25 identical OTUs) and Chitinophagaceae (8 out of 11 identical OTUs). In addition to this, two genera were found solely present in the low LUI core: SMB53 (555,945) and Labrys (543,156). OTU 543,156, however, was found in the core bacterial community composition of low LUI rhizosphere, too.

4. Discussion

4.1. The Role of LUI for Bacterial Diversity at the Plant–Soil Interface at Different Stages of Plant Development

In this study, we compared the putative core bacterial communities of the agricultural important grass *Dactylis glomerata* L. under the influence of different LUIs. Plants were sampled at three different dates from sites exposed to high and low LUI, respectively, to characterize the bacterial community composition in the endosphere and rhizosphere as well as in the bulk soil.

The analysis of α-diversity showed an increase of microbial diversity under high LUI within all plant stages only in the bulk soil. This might be explained by a more transient and patchy bacterial colonization under higher LUI [33]. In contrast, under low land use intensity, bacterial communities are differing strongly between the measured timepoints during the season, indicating a clear fluctuation of bacterial communities over the season, which is not the case under high land use intensity. Lower α-diversity as a result of low LUI, which we attribute to less disturbance, results in a more stable and consistent composition of bacterial communities [34]. However, we could neither find a significant impact of LUI on the bacterial α-diversity in the rhizosphere, nor in the root endosphere. This is in accordance with several studies showing that plant species have a stronger effect on plant-associated microbial communities than soil parameters [35,36].

Analysis of β-diversity within the root endosphere revealed a significant impact of land use intensity in June, indicating a difference in the presence and/or absence of certain OTUs. This might be due to seed production in June, which generates high metabolic expenses for the plant, which are caused by the synthesis of storage products (including proteins, starch and lipids), the uptake of mineral nutrients, as well as nutrient translocation from the site of synthesis to seed assimilation [37]. Thus, mainly at sites with low land use intensity and reduced amounts of plant available nitrogen, it is essential for optimal performance of the plant to recruit bacteria, which are capable of forming plant available nitrogen either by mineralization of dead biomass or by nitrogen fixation, which might explain differences in β-diversity in June between sites with different land use intensities.

As expected, a significant difference between LUIs was also observed in the rhizosphere and bulk soil at every sampling date. In addition, we also found a significant difference in the presence and/or absence of certain OTUs (unweighted Unifrac) between different sampling dates, within high and low LUI, respectively, in all compartments (Table S8). Thus, the results suggest a root-associated bacterial community composition that is selected by the plant throughout the season, which is in accordance with numerous studies [2,38,39] and additional effects driven by LUI, which differ in their degree and direction towards shifts in the abundance of single OTUs dependent on the sampling date. As the LUI index we used in our study takes into account various parameters, which are affected by land use intensity, including fertilization and subsequent changes of nitrogen pools in soils, the observed LUI-mediated effects might be coupled to altered soil parameters, which in turn affected the bacterial community composition.

4.2. The LUI Independent Bacterial Core Microbiome of D. glomerate

The main focus of this study was to identify a putative plant-associated core bacterial community of *D. glomerata* in different plant compartments, because bacteria that are consistently found across samples subjected to different conditions likely provide critical ecological functions. Indeed, we found plant-associated bacterial communities that were shared across all sampling seasons under both LUIs (Figure 2). *Pseudomonas veronii* was found in the root core of both LUIs, emphasizing its importance for *D. glomerata*. Previous studies have demonstrated that root-associated *P. veronii* exhibits high biocontrol potential by increasing the bioavailability of phosphate and ammonia in the soil [40] and by synthesizing indole-3-acetic acid (IAA), which is of major importance, since it stimulates cell elongation and cell division of the plant [41]. Furthermore, a high nematodical activity was observed within this species [42]. It was shown that the abundance of plant-parasitic nematodes is significantly increased in *D. glomerata* compared to other grasses and legumes [43]. This might be critical for *D. glomerata* since it can affect the competitive ability of the plant significantly. Moreover, other *Pseudomonas* spp. were reported to possess a wide variety of plant growth promoting traits and beneficial properties. These include the production of various phytohormones like IAA, cytokinins or gibberillins, as well as nitrogen fixation and production of antimicrobial compounds [44]. Additionally, *Rhizobium* and *Bradyrhizobium* were found in the plant-associated core bacterial composition independent of LUI. Both genera are known for their nitrogen-fixing abilities within nodules of leguminous plants [45,46] and for the production of phytohormones [47,48]. Moreover, genes for nitrogen fixation of the endophytic

Rhizobium spp. were found in high abundances within tissues of the perennial grass sugarcane, indicating their particular importance for plant-associated nitrogen fixation in perennial grasses [49]. Due to its high abundance in roots, it is presumably a key player for plant health.

In general, more OTUs were found in total and relative numbers in the core bacterial communities (shared across all seasons, high vs. low LUI) of the rhizosphere compared to roots, which seems reasonable, as there is no obstacle to overcome like passing the cell wall of the plant. Interestingly, the most prominent genus found in root core bacterial communities (*Pseudomonas*) was not as ubiquitously found in the rhizosphere core under both LUIs. This might be due to higher competition and the strong adaptation of *Pseudomonas* spp. to the specific conditions present in the root interior. The majority of genera found in the rhizospheric cores (high and low LUI) were *Rhodoplanes*, *Methylibium*, *Kaistobacter* and *Bradyrhizobium*, which were frequently isolated from other rhizosphere environments [50–53]. Among those, *Bradyrhizobium* is the only one that was also found in the root core bacterial communities under both LUIs, highlighting its importance for the plant. While representatives of *Rhodoplanes* have been characterized as a facultative photo-organotroph and potentially nitrate-fixing bacteria [54,55], *Methylibium* was described as a facultative methylotroph that actively utilizes root exudates [51]. Furthermore, species within the latter are involved in the degradation of aromatic hydrocarbon and methyl tert-butyl ether [50]. Members of *Kaistobacter* have also been reported to be involved in the degradation of aromatic compounds and suggested to suppress bacterial wilt disease [56].

4.3. Microbial Variation of Core OTUs as Influenced by LUI

In addition to the LUI-independent plant-associated core bacterial community composition (shared throughout all seasons and both LUIs), we also found taxa that were only present in the core under high or low LUI (shared throughout all seasons, but only on high or low LUI). In roots, the OTU assigned to *Caulobacter* was present exclusively in all samples from low LUI sites, constituting part of the core bacterial communities in roots. Microbes within this genus were reported to produce IAA and solubilize inorganic phosphate [57]. A study on bacterial communities in different grapevine cultivars showed higher abundance of *Caulobacter* under organic production versus integrated pest management [58], which would be in line with our findings of *Caulobacter* under low management intensity. In turn, a single OTU assigned to *Labrys* was found exclusively in the core of sites subjected to high LUI (shared throughout all seasons on high LUI). Species within this genus are frequently isolated from various soil and sediment samples but were also found in the rhizosphere of Korean ginseng [59–61], and as an endophyte in *Clerodendrum colebrookianum* [62]. Though no strain within this genus has been described in the context of plant growth-promoting traits, numerous species of *Labrys* isolated from sediment and soil were shown to possess the ability to reduce nitrate, assimilate various amino acids and sugars, have catalase activity etc. [59,60], which might also play an important role in supporting the plant under a high amount of available nitrate. *Agrobacterium* (e.g., *Agrobacterium tumefaciens*) is widely known for its ability to transmit plasmid T-DNA into plant cells [3]. Furthermore, several *Agrobacterium* species carry pathogenic capacity on Ti (tumorigenic) or Ri (rhizogenic) plasmids, which can cause the induction of tumor-like growth and reduce seed production, e.g., crown-gall or hairy root disease [63]. On the other hand, various nonpathogenic *Agrobacterium* spp. have been found lacking those plasmids [64]. Indeed, numerous *Agrobacterium* spp. appeared to contribute to plant growth by phosphate solubilization, nitrogen fixation and siderophore production [65].

We found several differences in the composition of core bacterial communities in the rhizosphere between high and low LUI, e.g., species within *Variovorax* were only found in the core of low LUI (shared throughout all seasons). *Variovorax paradoxus* is frequently described as the plant-growth promoting genus, including traits like the reduction of plant stress, increasing nutrient availability and inhibiting growth of pathogens by degrading N-acyl homoserine-lactones. The latter constitute mechanisms related to their catabolic capacities [65–67], which are likely critical to satisfy the metabolic expenses of the plant as they are found in the core microbiome of low LUI. Furthermore, *Sphingomonas wittichii* and *Bosea genosp* were only found in the low LUI core (shared throughout all seasons). Numerous studies showed their plant-growth promoting abilities due to their production of phytohormones [68,69]. Moreover, other species found only in the low LUI core within the genera *Sphingomonas* and *Bosea* were shown to be diazotrophic, solubilize inorganic phosphorus and are involved in biocontrol [70]. *Asticcacaulis biprosthecium* has recently been found in the rhizosphere of maize, however, their plant-growth promoting abilities have not been further described [71].

Genera exclusively found under high LUI, but shared throughout all sampling seasons were *Rhizobium*, *Mycoplana*, *Labrys*, *Adhaeribacter*, *Paucibacter* and *Microlunatus*, all of which have been described to harbor plant-growth promoting functions and thus enhancing plant performance [72–77].

OTUs found in the rhizosphere and bulk samples differed as expected, but all taxa that were detected in the bulk soil core bacterial communities were also found in the rhizosphere core under both LUIs. Differences in the core of bulk soil and the rhizosphere may be due to the selective attraction of taxa by the plant. Interestingly, LUI influenced the proportion of OTUs that contributed to the uniquely found OTUs (shared among all sampling seasons, but only on high or low LUI) and on the number of taxa in the rhizosphere (total OTUs found throughout all seasons), as OTUs were assigned to more different taxa under high LUI. Though the prevalent taxa specific for one plant stage were comparable in the rhizosphere, the amount and proportion of OTUs contributing to the core were lower under low LUI (156/19.4%) compared to high LUI (235/25.35%). Furthermore, the higher amount of OTUs that were present in the low LUI associated core bacterial communities compared to high LUI indicates a higher variability of bacteria colonizing the rhizosphere. In a low nutrient environment, specific recruitment of microorganisms is crucial to the plant to enhance plant fitness and growth [78]. Thus, a lower amount of similar genera may be found, indicating a more selective attraction of soil bacterial communities under low LUI by plant exudation throughout the season. Furthermore, the observation that seven OTUs were present in root samples and 124 in rhizosphere samples independent of LUI and sampling timepoints imply that these taxa are highly persistent and ubiquitous in agricultural soil.

5. Conclusions

During the last years, several studies investigated the influence of land management on plant-associated bacteria, thereby focusing on single plant development stages or compartments. As frequently occurring plant-associated bacterial assemblages are presumably critical for plant fitness [18], the identification of a plant-associated core bacterial community composition might be the first step of defining a "healthy" bacterial community to unravel the ecology of plant-associated bacterial consortia and predict community responses to environmental changes. Up to date, this is the first study defining the composition of a stable plant-associated core bacterial community composition in the rhizosphere and endosphere as well as its dependency on LUI for a plant species of agricultural importance. We found the genera *Pseudomonas*, *Rhizobium* and *Bradyrhizobium* to be part of the consistent root core plant-associated bacteria of *D. glomerata* independent of LUI and stable along the season. The majority of genera identified as the core bacterial communities of the rhizosphere compartment belonged to *Rhodoplanes*, *Methylibium*, *Kaistobacter* and *Bradyrhizobium*. Their persistent occurrence independent of LUI or the growth stage as well as their plant-growth promoting traits supporting plant health could be a first insight into the composition of a life-supporting essential part of the plant microbiome. Nevertheless, we could not infer from the measured co-occurrence of plant-associated bacterial communities across different LUI how the detected species react and interact with their host plants, although many of the identified taxa have been associated with plant growth promoting traits in previous studies [79]. In this context, also the co-occurrence of fungi and bacteria especially in the plants' rhizosphere compartment should be assessed to account for the presence and contribution of fungi in plant-associated microbiomes.

A higher compositional variation throughout different sampling dates was observed under low LUI compared to high LUI, suggesting a stronger adaption of plant-associated bacteria under low LUI. Taking the minimum cut-off of 0.001% (i.e., required number of reads for one OTU to be kept was 103) after cut-off into account, these results were significant.

Taking the functions associated to the detected bacterial species into account, we could speculate that the major functionality of the plant-associated core bacterial community might be linked in particular to the plant growth and stress response. However, to obtain deeper insights into the functionality of a "healthy" core bacterial microbiome and its resilience to disturbances, it is of importance to further analyze the functional traits of the identified microbial communities to reveal their functional potential. This will gain deeper knowledge on core plant-associated bacteria of a healthy versus a diseased state of the plant and the predictability of the fitness state of the plant to improve plant performance in agricultural production systems. Though our study revealed the existence of shared plant-associated bacterial taxa under different land use intensities, these findings were specific for *D. glomerata* in the investigated grassland soils. Thus, it would be tempting to assess also bacterial communities associated with other *D. glomerata* cultivars or to study broader biogeographical patterns of the root associated microbiome of *D. glomerata* or other grass species of Poaceae.

Supplementary Materials: The following materials are available online at http://www.mdpi.com/1424-2818/12/10/392/s1, Figure S1: Rarefaction curves, Table S1: Summary of sampling plot parameters, Table S2: Summary of edaphic parameters, Table S3: The impact of LUI on the β-diversity within one sampling season in all compartments (between sample diversity), Table S4: The impact of sampling season on the β-diversity within one LUI in all compartments (between sample diversity), Table S5: Relative abundance of assigned phyla, Table S6: Relative abundance of assigned genera, Table S7: Classification of core OTUs in roots and high LUI. Shared core OTUs (May, June and October) high LUI: 12, Table S8: Classification of core OTUs in roots and low LUI. Shared core OTUs (May, June and October) low LUI: 14, Table S9: Classification of shared core OTUs that were found in 95% in high LUI samples across all sampling seasons (rhizosphere), Table S10: Classification of shared core OTUs that were found in 95% in low LUI samples across all sampling seasons (rhizosphere), Table S11: Classification of shared core OTUs that were found in 95% in high LUI samples across all sampling seasons (bulk soil), Table S12: Classification of shared core OTUs that were found in 95% in high LUI samples across all sampling seasons (bulk soil).

Author Contributions: Conceptualization, M.C.R., J.J., P.S. and M.S.; Methodology, J.E. and G.V.; Software, G.V.; Investigation, J.E.; Writing—Original Draft Preparation, J.E. and B.S., Writing—Review and Editing, B.S., S.S., M.C.R., J.J., P.S. and M.S.; Supervision S.S., P.S., M.S., Funding Acquisition, M.S. All authors have read and agreed to the published version of the manuscript.

Funding: This research was (partly) funded by DFG Priority Program 1374, "Infrastructure-Biodiversity-Exploratories".

Acknowledgments: We thank Ingo Schöning for providing data concerning soil parameters. We thank the managers of the three Exploratories, Kirsten Reichel-Jung, Katrin Lorenzen, Martin Gorke and all former managers for their work in maintaining the plot and project infrastructure; Christiane Fischer for giving support through the central office, Michael Owonibi for managing the central data base, and Markus Fischer, Eduard Linsenmair, Dominik Hessenmöller, Daniel Prati, Ingo Schöning, François Buscot, Ernst-Detlef Schulze, Wolfgang W. Weisser and the late Elisabeth Kalko for their role in setting up the Biodiversity Exploratories project. Field work permits were issued by the responsible state environmental offices of Baden-Württemberg, Thüringen, and Brandenburg.

References

1. Martínez-Viveros, O.; Jorquera, M.A.; Crowley, D.E.; Gajardo, G.; Mora, M.L. Mechanisms and practical considerations involved in plant growth promotion by rhizobacteria. *J. Soil. Sci. Plant. Nutr.* **2010**, *10*, 293–319. [CrossRef]

2. Berendsen, R.L.; Pieterse, C.M.J.; Bakker, P.A.H.M. The rhizosphere microbiome and plant health. *Trends Plant. Sci.* **2012**, *17*, 478–486. [CrossRef] [PubMed]

3. Gaiero, J.; McCall, C.; Thompson, K.; Day, N.; Best, A.; Dunfield, K. Inside the root microbiome: Bacterial root endophytes and plant growth promotion. *Am. J. Bot.* **2013**, *100*. [CrossRef]

4. Reinhold-Hurek, B.; Bünger, W.; Burbano, C.S.; Sabale, M.; Hurek, T. Roots shaping their microbiome: Global hotspots for microbial activity. *Annu. Rev. Phytopathol.* **2015**, *53*, 403–424. [CrossRef] [PubMed]

5. Turner, T.R.; James, E.K.; Poole, P.S. The plant microbiome. *Genome. Biol.* **2013**, *14*, 209. [CrossRef] [PubMed]

6. Lugtenberg, B.; Kamilova, F. Plant-Growth-Promoting Rhizobacteria. *Annu. Rev. Microbiol.* **2009**, *63*, 541–556. [CrossRef] [PubMed]

7. Reinhold-Hurek, B.; Hurek, T. Living inside plants: Bacterial endophytes. *Curr. Opin. Plant. Biol.* **2011**, *14*, 435–443. [CrossRef]

8. Truyens, S.; Weyens, N.; Cuypers, A.; Vangronsveld, J. Bacterial seed endophytes: Genera, vertical transmission and interaction with plants. *Environ. Microbiol. Rep.* **2015**, *7*, 40–50. [CrossRef]

9. Helliwell, J.R.; Sturrock, C.J.; Miller, A.J.; Whalley, W.R.; Mooney, S.J. The role of plant species and soil condition in the structural development of the rhizosphere. *Plant. Cell Environ.* **2019**, *42*, 1974–1986. [CrossRef]

10. Kowalchuk, G.A.; Buma, D.S.; de Boer, W.; Klinkhamer, P.G.L.; van Veen, J.A. Effects of above-ground plant species composition and diversity on the diversity of soil-borne microorganisms. *Antonie Van Leeuwenhoek* **2002**, *81*, 509. [CrossRef]

11. Compant, S.; Reiter, B.; Sessitsch, A.; Nowak, J.; Clément, C.; Ait Barka, E. Endophytic Colonization of *Vitis vinifera* L. by Plant Growth-Promoting Bacterium *Burkholderia* sp. Strain PsJN. *Appl. Environ. Microbiol.* **2005**, *71*, 1685–1693. [CrossRef] [PubMed]

12. Tuteja, N.; Sopory, S.K. Chemical signaling under abiotic stress environment in plants. *Plant. Signal. Behav.* **2008**, *3*, 525–536. [CrossRef] [PubMed]

13. Ernebjerg, M.; Kishony, R. Distinct Growth Strategies of Soil Bacteria as Revealed by Large-Scale Colony Tracking. *Appl. Environ. Microbiol.* **2012**, *78*, 1345–1352. [CrossRef] [PubMed]

14. Tardieu, F. Plant response to environmental conditions: Assessing potential production, water demand, and negative effects of water deficit. *Front. Physiol.* **2013**, *4*. [CrossRef] [PubMed]

15. Estendorfer, J.; Stempfhuber, B.; Haury, P.; Vestergaard, G.; Rillig, M.C.; Joshi, J.; Schröder, P.; Schloter, M. The Influence of Land Use Intensity on the Plant-Associated Microbiome of *Dactylis glomerata* L. *Front. Plant. Sci.* **2017**, *8*. [CrossRef] [PubMed]

16. Lundberg, D.; Lebeis, S.; Herrera Paredes, S.; Yourstone, S.; Gehring, J.; Malfatti, S.; Tremblay, J.; Engelbrektson, A.; Kunin, V.; Rio, T.; et al. Defining the core Arabidopsis thaliana root microbiome. *Nature* **2012**, *488*, 86–90. [CrossRef]

17. Pfeiffer, S.; Mitter, B.; Oswald, A.; Schloter-Hai, B.; Schloter, M.; Declerck, S.; Sessitsch, A. Rhizosphere microbiomes of potato cultivated in the High Andes show stable and dynamic core microbiomes with different responses to plant development. *FEMS Microbiol. Ecol.* **2017**, *93*. [CrossRef]

18. Bowen, J.L.; Kearns, P.J.; Byrnes, J.E.K.; Wigginton, S.; Allen, W.J.; Greenwood, M.; Tran, K.; Yu, J.; Cronin, J.T.; Meyerson, L.A. Lineage overwhelms environmental conditions in determining rhizosphere bacterial community structure in a cosmopolitan invasive plant. *Nat. Commun.* **2017**, *8*, 433. [CrossRef]

19. Bluthgen, N.; Dormann, C.F.; Prati, D.; Klaus, V.H.; Kleinebecker, T.; Holzel, N.; Alt, F.; Boch, S.; Gockel, S.; Hemp, A.; et al. A quantitative index of land-use intensity in grasslands: Integrating mowing, grazing and fertilization. *Basic Appl. Ecol.* **2012**, *13*, 207–220. [CrossRef]

20. Houba, V.J.G.; Temminghoff, E.J.M.; Gaikhorst, G.A.; van Vark, W. Soil analysis procedures using 0.01 M calcium chloride as extraction reagent. *Commun. Soil. Sci. Plant. Anal.* **2000**, *31*, 1299–1396. [CrossRef]

21. Lueders, T.; Manefield, M.; Friedrich, M.W. Enhanced sensitivity of DNA- and rRNA-based stable isotope probing by fractionation and quantitative analysis of isopycnic centrifugation gradients. *Environ. Microbiol.* **2004**, *6*, 73–78. [CrossRef] [PubMed]

22. Dorn-In, S.; Bassitta, R.; Schwaiger, K.; Bauer, J.; Hölzel, C.S. Specific amplification of bacterial DNA by optimized so-called universal bacterial primers in samples rich of plant DNA. *J. Microbiol. Methods* **2015**, *113*, 50–56. [CrossRef] [PubMed]

23. Caporaso, J.G.; Kuczynski, J.; Stombaugh, J.; Bittinger, K.; Bushman, F.D.; Costello, E.K.; Fierer, N.; Peña, A.G.; Goodrich, J.K.; Gordon, J.I.; et al. QIIME allows analysis of high-throughput community sequencing data. *Nat. Methods* **2010**, *7*, 335–336. [CrossRef] [PubMed]

24. Schubert, M.; Lindgreen, S.; Orlando, L. AdapterRemoval v2: Rapid adapter trimming, identification, and read merging. *BMC Res. Notes* **2016**, *9*, 88. [CrossRef] [PubMed]

25. Schmieder, R.; Edwards, R. Fast identification and removal of sequence contamination from genomic and metagenomic datasets. *PLoS ONE* **2011**, *6*, e17288. [CrossRef] [PubMed]

26. DeSantis, T.Z.; Hugenholtz, P.; Larsen, N.; Rojas, M.; Brodie, E.L.; Keller, K.; Huber, T.; Dalevi, D.; Hu, P.; Andersen, G.L. Greengenes, a chimera-checked 16S rRNA gene database and workbench compatible with ARB. *Appl. Environ. Microbiol.* **2006**, *72*, 5069–5072. [CrossRef] [PubMed]

27. Wang, Q.; Garrity, G.M.; Tiedje, J.M.; Cole, J.R. Naïve Bayesian Classifier for Rapid Assignment of rRNA Sequences into the New Bacterial Taxonomy. *Appl. Environ. Microbiol.* **2007**, *73*, 5261–5267. [CrossRef]

28. Chao, A. Nonparametric Estimation of the Number of Classes in a Population. *Scand. Stat. Theory Appl.* **1984**, *11*, 265–270.

29. Southwood, T.R.E.; Henderson, P.A. *Ecological Methods*, 3rd ed.; Blackwell Science Ltd.: Hoboken, NJ, USA, 2000; p. 592.

30. Lozupone, C.A.; Hamady, M.; Kelley, S.T.; Knight, R. Quantitative and Qualitative β Diversity Measures Lead to Different Insights into Factors That Structure Microbial Communities. *Appl. Environ. Microbiol.* **2007**, *73*, 1576–1585. [CrossRef]

31. Oksanen, F.J.; Blanchet, F.G.; Friendly, M.; Kindt, R.; Legendre, P.; McGlinn, D.; Minchin, P.; O'Hara, R.B.; Simpson, G.; Solymos, P.; et al. Vegan: Community Ecology Package. Available online: https://cran.r-project.org/src/contrib/Archive/vegan/ (accessed on 10 August 2017).

32. Calculate and Draw Custom Venn Diagrams. Available online: http://bioinformatics.psb.ugent.be/webtools/Venn/ (accessed on 15 January 2018).

33. Garbeva, P.; van Elsas, J.D.; van Veen, J.A. Rhizosphere microbial community and its response to plant species and soil history. *Plant. Soil.* **2008**, *302*, 19–32. [CrossRef]

34. Bevivino, A.; Paganin, P.; Bacci, G.; Florio, A.; Pellicer, M.S.; Papaleo, M.C.; Mengoni, A.; Ledda, L.; Fani, R.; Benedetti, A.; et al. Soil bacterial community response to differences in agricultural management along with seasonal changes in a Mediterranean region. *PLoS ONE* **2014**, *9*, e105515. [CrossRef] [PubMed]

35. Grayston, S.J.; Wang, S.; Campbell, C.D.; Edwards, A.C. Selective influence of plant species on microbial diversity in the rhizosphere. *Soil. Biol. Biochem.* **1998**, *30*, 369–378. [CrossRef]

36. Gottel, N.R.; Castro, H.F.; Kerley, M.; Yang, Z.; Pelletier, D.A.; Podar, M.; Karpinets, T.; Uberbacher, E.; Tuskan, G.A.; Vilgalys, R.; et al. Distinct microbial communities within the endosphere and rhizosphere of Populus deltoides roots across contrasting soil types. *Appl. Environ. Microbiol.* **2011**, *77*, 5934–5944. [CrossRef] [PubMed]

37. Munier-Jolain, N.; Salon, C. Are the carbon costs of seed production related to the quantitative and qualitative performance? An appraisal of legumes and other crops. *Plant. Cell Environ.* **2005**, *28*, 1388–1395. [CrossRef]

38. Berg, G.; Grube, M.; Schloter, M.; Smalla, K. Unraveling the plant microbiome: Looking back and future perspectives. *Front. Microbiol.* **2014**, *5*. [CrossRef]

39. Lareen, A.; Burton, F.; Schäfer, P. Plant root-microbe communication in shaping root microbiomes. *Plant. Mol. Biol.* **2016**, *90*, 575–587. [CrossRef]

40. Montes, C.; Altimira, F.; Canchignia, H.; Castro, Á.; Sánchez, E.; Miccono, M.; Tapia, E.; Sequeida, Á.; Valdés, J.; Tapia, P.; et al. A draft genome sequence of Pseudomonas veronii R4: A grapevine (*Vitis vinifera* L.) root-associated strain with high biocontrol potential. *Stand. Genom. Sci.* **2016**, *11*, 76. [CrossRef]

41. Perrot-Rechenmann, C. Cellular responses to auxin: Division versus expansion. *Cold Spring Harb. Perspect. Biol.* **2010**, *2*, a001446. [CrossRef]

42. Canchignia, H.; Altimira, F.; Montes, C.; Sánchez, E.; Tapia, E.; Miccono, M.; Espinoza, D.; Aguirre, C.; Seeger, M.; Prieto, H. Candidate nematicidal proteins in a new Pseudomonas veronii isolate identified by its antagonistic properties against Xiphinema index. *J. Gen. Appl. Microbiol.* **2017**, *63*, 11–21. [CrossRef]

43. Viketoft, M.; Palmborg, C.; Sohlenius, B.; Huss-Danell, K.; Bengtsson, J. Plant species effects on soil nematode communities in experimental grasslands. *Appl. Soil Ecol.* **2005**, *30*, 90–103. [CrossRef]

44. Oteino, N.; Lally, R.D.; Kiwanuka, S.; Lloyd, A.; Ryan, D.; Germaine, K.J.; Dowling, D.N. Plant growth promotion induced by phosphate solubilizing endophytic Pseudomonas isolates. *Front. Microbiol.* **2015**, *6*, 745. [CrossRef] [PubMed]

45. Lodwig, E.M.; Hosie, A.H.F.; Bourdès, A.; Findlay, K.; Allaway, D.; Karunakaran, R.; Downie, J.A.; Poole, P.S. Amino-acid cycling drives nitrogen fixation in the legume–Rhizobium symbiosis. *Nature* **2003**, *422*, 722–726. [CrossRef] [PubMed]

46. Fujita, H.; Aoki, S.; Kawaguchi, M. Evolutionary Dynamics of Nitrogen Fixation in the Legume–Rhizobia Symbiosis. *PLoS ONE* **2014**, *9*, e93670. [CrossRef]

47. Chi, F.; Shen, S.-H.; Cheng, H.-P.; Jing, Y.-X.; Yanni, Y.G.; Dazzo, F.B. Ascending migration of endophytic rhizobia, from roots to leaves, inside rice plants and assessment of benefits to rice growth physiology. *Appl. Environ. Microbiol.* **2005**, *71*, 7271–7278. [CrossRef] [PubMed]

48. Egamberdieva, D.; Wirth, S.J.; Alqarawi, A.A.; Abd_Allah, E.F.; Hashem, A. Phytohormones and Beneficial Microbes: Essential Components for Plants to Balance Stress and Fitness. *Front. Microbiol.* **2017**, *8*. [CrossRef] [PubMed]

49. Fischer, D.; Pfitzner, B.; Schmid, M.; Simões-Araújo, J.L.; Reis, V.M.; Pereira, W.; Ormeño-Orrillo, E.; Hai, B.; Hofmann, A.; Schloter, M.; et al. Molecular characterisation of the diazotrophic bacterial community in uninoculated and inoculated field-grown sugarcane (*Saccharum* sp.). *Plant. Soil* **2012**, *356*, 83–99. [CrossRef]

50. Nakatsu, C.H.; Hristova, K.; Hanada, S.; Meng, X.-Y.; Hanson, J.R.; Scow, K.M.; Kamagata, Y. Methylibium petroleiphilum gen. nov., sp. nov., a novel methyl tert-butyl ether-degrading methylotroph of the Betaproteobacteria. *Int. J. Syst. Evol. Microbiol.* **2006**, *56*, 983–989. [CrossRef]

51. Mao, Y.; Li, X.; Smyth, E.M.; Yannarell, A.C.; Mackie, R.I. Enrichment of specific bacterial and eukaryotic microbes in the rhizosphere of switchgrass (*Panicum virgatum* L.) through root exudates. *Environ. Microbiol. Rep.* **2014**, *6*, 293–306. [CrossRef]

52. Rouws, L.; Leite, J.; Feitosa de Matos, G.; Zilli, J.; Reed Rodrigues Coelho, M.; Xavier, G.; Fischer, D.; Hartmann, A.; Reis, V.; Baldani, J. Endophytic Bradyrhizobium spp. isolates from sugarcane obtained through different culture strategies. *Environ. Microbiol. Rep.* **2013**, *6*. [CrossRef]

53. Gkarmiri, K.; Mahmood, S.; Ekblad, A.; Alström, S.; Högberg, N.; Finlay, R. Identifying the Active Microbiome Associated with Roots and Rhizosphere Soil of Oilseed Rape. *Appl. Environ. Microbiol.* **2017**, *83*, e01938-17. [CrossRef]

54. Hiraishi, A.; Ueda, Y. Rhodoplanes gen. nov., a New Genus of Phototrophic Bacteria Including Rhodopseudomonas rosea as Rhodoplanes roseus comb. nov. and Rhodoplanes elegans sp. nov. *Int. J. Syst. Evol. Microbiol.* **1994**, *44*, 665–673. [CrossRef]

55. Buckley, D.H.; Huangyutitham, V.; Hsu, S.-F.; Nelson, T.A. Stable Isotope Probing with $^{15}N_2$ Reveals Novel Noncultivated Diazotrophs in Soil. *Appl. Environ. Microbiol.* **2007**, *73*, 3196–3204. [CrossRef]

56. Liu, X.; Zhang, S.; Jiang, Q.; Bai, Y.; Shen, G.; Li, S.; Ding, W. Using community analysis to explore bacterial indicators for disease suppression of tobacco bacterial wilt. *Sci. Rep.* **2016**, *6*, 36773. [CrossRef] [PubMed]

57. Chimwamurombe, P.M.; Grönemeyer, J.L.; Reinhold-Hurek, B. Isolation and characterization of culturable seed-associated bacterial endophytes from gnotobiotically grown Marama bean seedlings. *FEMS Microbiol. Ecol.* **2016**, *92*, fiw083. [CrossRef]

58. Campisano, A.; Antonielli, L.; Pancher, M.; Yousaf, S.; Pindo, M.; Pertot, I. Bacterial Endophytic Communities in the Grapevine Depend on Pest Management. *PLoS ONE* **2014**, *9*, e112763. [CrossRef] [PubMed]

59. Miller, J.A.; Kalyuzhnaya, M.G.; Noyes, E.; Lara, J.C.; Lidstrom, M.E.; Chistoserdova, L. Labrys methylaminiphilus sp. nov., a novel facultatively methylotrophic bacterium from a freshwater lake sediment. *Int. J. Syst. Evol. Microbiol.* **2005**, *55*, 1247–1253. [CrossRef] [PubMed]

60. Carvalho, M.; De Marco, P.; Duque, A.; Pacheco, C.; Janssen, D.; Castro, P. Labrys portucalensis sp. nov., a fluorobenzene-degrading bacterium isolated from an industrially contaminated sediment in northern Portugal. *Int. J. Syst. Evol. Microbiol.* **2008**, *58*, 692–698. [CrossRef]

61. Nguyen, N.-L.; Kim, Y.-J.; Hoang, V.-A.; Kang, J.-P.; Wang, C.; Zhang, J.; Kang, C.-H.; Yang, D.-C. Labrys soli sp. nov., isolated from the rhizosphere of ginseng. *Int. J. Syst. Evol. Microbiol.* **2015**, *65*, 3913–3919. [CrossRef]

62. Passari, A.K.; Mishra, V.K.; Leo, V.V.; Gupta, V.K.; Singh, B.P. Phytohormone production endowed with antagonistic potential and plant growth promoting abilities of culturable endophytic bacteria isolated from Clerodendrum colebrookianum Walp. *Microbiol. Res.* **2016**, *193*, 57–73. [CrossRef]

63. Chihaoui, S.A.; Trabelsi, D.; Jdey, A.; Mhadhbi, H.; Mhamdi, R. Inoculation of Phaseolus vulgaris with the nodule-endophyte *Agrobacterium* sp. 10C2 affects richness and structure of rhizosphere bacterial communities and enhances nodulation and growth. *Arch. Microbiol.* **2015**, *197*, 805–813. [CrossRef]

64. Bosmans, L.; Moerkens, R.; Wittemans, L.; De Mot, R.; Rediers, H.; Lievens, B. Rhizogenic agrobacteria in hydroponic crops: Epidemics, diagnostics and control. *Plant. Pathol.* **2017**, *66*, 1043–1053. [CrossRef]

65. Pereira, S.I.; Castro, P.M. Diversity and characterization of culturable bacterial endophytes from Zea mays and their potential as plant growth-promoting agents in metal-degraded soils. *Environ. Sci. Pollut. Res.* **2014**, *21*, 14110–14123. [CrossRef]

66. Leadbetter, J.R.; Greenberg, E.P. Metabolism of Acyl-Homoserine Lactone Quorum-Sensing Signals by *Variovorax paradoxus*. *J. Bacteriol.* **2000**, *182*, 6921–6926. [CrossRef] [PubMed]

67. Han, J.-I.; Choi, H.-K.; Lee, S.-W.; Orwin, P.M.; Kim, J.; LaRoe, S.L.; Kim, T.-g.; O'Neil, J.; Leadbetter, J.R.; Lee, S.Y.; et al. Complete Genome Sequence of the Metabolically Versatile Plant Growth-Promoting Endophyte *Variovorax paradoxus* S110. *J. Bacteriol.* **2011**, *193*, 1183–1190. [CrossRef] [PubMed]

68. Shen, S.Y.; Fulthorpe, R. Seasonal variation of bacterial endophytes in urban trees. *Front. Microbiol.* **2015**, *6*, 427. [CrossRef]

69. Khan, A.L.; Waqas, M.; Asaf, S.; Kamran, M.; Shahzad, R.; Bilal, S.; Khan, M.A.; Kang, S.-M.; Kim, Y.-H.; Yun, B.-W.; et al. Plant growth-promoting endophyte Sphingomonas sp. LK11 alleviates salinity stress in Solanum pimpinellifolium. *Environ. Exp. Bot.* **2017**, *133*, 58–69. [CrossRef]

70. Correa-Galeote, D.; Bedmar, E.J.; Arone, G.J. Maize Endophytic Bacterial Diversity as Affected by Soil Cultivation History. *Front. Microbiol.* **2018**, *9*. [CrossRef] [PubMed]

71. Walters, W.A.; Jin, Z.; Youngblut, N.; Wallace, J.G.; Sutter, J.; Zhang, W.; González-Peña, A.; Peiffer, J.; Koren, O.; Shi, Q.; et al. Large-scale replicated field study of maize rhizosphere identifies heritable microbes. *Proc. Natl. Acad. Sci. USA* **2018**, *115*, 7368–7373. [CrossRef]

72. Dohrmann, A.; Küting, M.; Jünemann, S.; Jaenicke, S.; Schlueter, A.; Tebbe, C. Importance of rare taxa for bacterial diversity in the rhizosphere of Bt-and conventional maize varieties. *ISME J.* **2012**, *7*. [CrossRef]

73. Egamberdiyeva, D.; Höflich, G. Influence of growth-promoting bacteria on the growth of wheat in different soils and temperatures. *Soil. Biol. Biochem.* **2003**, *35*, 973–978. [CrossRef]

74. Fernández-González, A.J.; Martínez-Hidalgo, P.; Cobo-Díaz, J.F.; Villadas, P.J.; Martínez-Molina, E.; Toro, N.; Tringe, S.G.; Fernández-López, M. The rhizosphere microbiome of burned holm-oak: Potential role of the genus Arthrobacter in the recovery of burned soils. *Sci. Rep.* **2017**, *7*, 6008. [CrossRef] [PubMed]

75. Madigan, M.; Cox, S.S.; Stegeman, R.A. Nitrogen fixation and nitrogenase activities in members of the family Rhodospirillaceae. *J. Bacteriol.* **1984**, *157*, 73–78. [CrossRef]

76. Laranjo, M.; Alexandre, A.; Oliveira, S. Legume growth-promoting rhizobia: An overview on the Mesorhizobium genus. *Microbiol. Res.* **2014**, *169*, 2–17. [CrossRef] [PubMed]

77. Hardoim, P.R.; van Overbeek, L.S.; Berg, G.; Pirttilä, A.M.; Compant, S.; Campisano, A.; Döring, M.; Sessitsch, A. The Hidden World within Plants: Ecological and Evolutionary Considerations for Defining Functioning of Microbial Endophytes. *Microbiol. Mol. Biol. Rev. MMBR* **2015**, *79*, 293–320. [CrossRef] [PubMed]

78. Dreccer, M.F.; Chapman, S.C.; Rattey, A.R.; Neal, J.; Song, Y.; Christopher, J.J.; Reynolds, M. Developmental and growth controls of tillering and water-soluble carbohydrate accumulation in contrasting wheat (*Triticum aestivum* L.) genotypes: Can we dissect them? *J. Exp. Bot.* **2013**, *64*, 143–160. [CrossRef]

79. Fitzpatrick, C.R.; Salas-González, I.; Conway, J.M.; Finkel, O.M.; Gilbert, S.; Russ, D.; Pereira Lima Teixeira, P.J.; Dangl, J.L. The Plant Microbiome: From Ecology to Reduction and Beyond. *Annu. Rev. Microbiol.* **2020**, *74*, 81–100. [CrossRef]

Article

Urochloa Grasses Swap Nitrogen Source When Grown in Association with Legumes in Tropical Pastures

Daniel M. Villegas [1], Jaime Velasquez [2], Jacobo Arango [1], Karen Obregon [2], Idupulapati M. Rao [1], Gelber Rosas [2] and Astrid Oberson [3,*]

[1]   International Center for Tropical Agriculture (CIAT), Cali 763537, Colombia; d.m.villegas@cgiar.org (D.M.V.); j.arango@cgiar.org (J.A.); i.rao@cgiar.org (I.M.R.)
[2]   Research Group in rural Development (GIADER), Universidad de la Amazonia, Porvenir Campus, Florencia 180001, Colombia; j.velasquez@udla.edu.co (J.V.); k.obregon@udla.edu.co (K.O.); g.rosas@udla.edu.co (G.R.)
[3]   ETH Zurich, Institute of Agricultural Sciences, 8315 Lindau, Switzerland
*   Correspondence: astrid.oberson@usys.ethz.ch; Tel.: +41-52-354-91-32

Received: 19 August 2020; Accepted: 1 November 2020; Published: 5 November 2020

Abstract: The degradation of tropical pastures sown with introduced grasses (e.g., *Urochloa* spp.) has dramatic environmental and economic consequences in Latin America. Nitrogen (N) limitation to plant growth contributes to pasture degradation. The introduction of legumes in association with grasses has been proposed as a strategy to improve N supply via symbiotic N_2 fixation, but the fixed N input and N benefits for associated grasses have hardly been determined in farmers' pastures. We have carried out on-farm research in ten paired plots of grass-alone (GA) vs. grass-legume (GL) pastures. Measurements included soil properties, pasture productivity, and sources of plant N uptake using ^{15}N isotope natural abundance methods. The integration of legumes increased pasture biomass production by about 74%, while N uptake was improved by two-fold. The legumes derived about 80% of their N via symbiotic N_2 fixation. The isotopic signature of N of grasses in GA vs. GL pastures suggested that sources of grass N are affected by sward composition. Low values of $\delta^{15}N$ found in some grasses in GA pastures indicate that they depend, to some extent, on N from non-symbiotic N_2 fixation, while $\delta^{15}N$ signatures of grasses in GL pastures pointed to N transfer to grass from the associated legume. The role of different soil–plant processes such as biological nitrification inhibition (BNI), non-symbiotic N_2 fixation by GA pastures and legume–N transfer to grasses in GL pastures need to be further studied to provide a more comprehensive understanding of N sources supporting the growth of grasses in tropical pastures.

Keywords: biological nitrogen fixation; nitrogen concentration; nitrogen transfer; ^{15}N natural abundance

1. Introduction

Deforestation in the tropics has been estimated at about 0.74 million km^2 from 2000 to 2012 [1]. Nearly half of it occurred in the South American rainforest [1]. In the Amazon basin, most of the cleared land has been converted to pastures sown with introduced grasses (mostly *Urochloa* spp., formerly known as *Brachiaria*) for livestock production [2,3]. In Colombia alone, more than 8% of the remaining forest area has been lost since 2000, yielding one of the highest deforestation rates in South America [2,4].

The majority of tropical pastures exist in some stage of degradation [5]. Pasture degradation is understood as a marked reduction in livestock production due to a significant decrease in forage yield and nutritional quality, the invasion of non-palatable plant species leading to bare soil patches (thus increasing susceptibility to erosion), soil compaction, acidification and reduced soil microbial biomass [6–8]. This phenomenon has tremendous economic and ecological implications, as it leaves large areas of degraded land and promotes a trend of continuing deforestation [9]. As an example of economic implications, in Brazil, every year, about 8 million hectares of degraded pastures require considerable investment for renewal and/or recovery [3], with an estimated annual cost of USD 100 to 200 ha^{-1}, i.e., around USD 1 billion per year in total at country level [10].

The predominant soils of the deforested area in the Colombian Amazon are highly weathered Haplic Ferralsols and Haplic Acrisols [11]. These acid soils typically have low total and available phosphorus (P) contents. Overgrazing and reduced pools of available nitrogen (N) and P in soil are seen as the principal causes of pasture degradation [6,8,12–14]. Thus, grass-legume (GL) associations are an important alternative to grass-alone (GA) pastures because of N input from the legume through symbiotic N$_2$ fixation [15]. Examples of legume species sown in association with *Urochloa* grasses are *Pueraria phaseoloides*, *Arachis pintoi*, *Desmodium ovalifolium*, *Centrosema* spp. and *Stylosanthes* spp. Despite the aggressive growth of *Urochloa* grasses [15,16], stable *Urochloa*/legume associations are possible where grazing is appropriately managed [17,18]. Provided sufficient P supply, tropical forage legumes obtain at least 70% of their N through symbiotic N$_2$ fixation [19]. A legume proportion range from 20% to 45% of total pasture dry matter has been estimated to cover the N requirement of pasture growth in the tropics [20], similar to grass-clover swards in temperate climates [21]. GL associations also significantly improve animal (meat and milk) productivity [18,22]. Such improvements have been largely related to the production of more forage biomass (mainly in dry season) and of better quality. Mixed grass-legume diets provide highly digestible protein, less structural carbohydrates and therefore increase animal forage voluntary intake, efficiency of nutrient conversion and live weight gain more than diets based on N fertilized grass monocultures [23–25]. Whilst the agricultural and environmental benefits of integration of legumes in tropical pastures have repeatedly been shown in researcher-managed fields [18], the adoption of legumes by farmers has remained rather small.

In pastoral systems, the quantity and quality of plant litter inputs are crucial for nutrient cycling [26]. In addition to aboveground litter, belowground inputs composed by root and rhizodeposition may enhance nutrient cycling and availability [27]. The belowground N input of clover growing in GL mixtures under temperate climate was around 40% of aboveground N [28] and about 50% of grass N was legume-derived [29]. Grasses growing in association with legumes thus benefit from symbiotically fixed N, with decomposing legume roots most likely being the main transfer pathway [29,30]. Preliminary research work on the determination of δ^{15}N values of shoot tissue as part of the study on pasture degradation [8] suggests that *Urochloa* spp. have different N acquisition strategies, resulting in N uptake from different sources, when growing alone than when growing with legumes. Indeed, *Urochloa* spp. have a variety of strategies, which could affect the δ^{15}N signature of their biomass [31]. First, grasses of the genus *Urochloa* can obtain 20% to 40% of their total N in the plant from the atmosphere through the association with N fixing bacteria (i.e., non-symbiotic N$_2$ fixation) [32,33]. Secondly, *U. humidicola* can suppress soil nitrification by releasing inhibitors from roots [34,35], which affects the δ^{15}N in plants [31]. Thus, the integration of legumes could advance toward improved N supply to the associated grass, either via the provision of fixed N$_2$, or via N sparing due to the reduced demand of mineralized soil N to support legume growth.

In temperate GL pastures, grasses and legumes mutually benefit to acquire N from diverse symbiotic and non-symbiotic sources, and transform N into biomass more efficiently than pure grass or legume fields [36]. Nevertheless, in spite of the fact that in tropical grasslands the beneficial effects of GL pastures on livestock productivity and soil fertility have been well studied, the underlying processes such as symbiotic N$_2$ fixation and N sources exploited by mixed pasture components have rarely been determined under farmers' pasture management conditions.

The objectives of this work carried out under conditions of farmers' practice, at the plot scale were to: (i) estimate the productivity and N uptake of GA vs. GL pastures; (ii) determine the symbiotic N$_2$ fixation by legumes; and (iii) evaluate the N sources of grasses growing alone vs. associated with legumes. We hypothesized that legumes would fix significant amounts of N, and that the associated grasses in GL pastures would benefit from the symbiotic N_2 fixation via legume N transfer, and that this legume N transfer would result in lower $\delta^{15}N$ value of grasses in GL than in GA pastures.

2. Materials and Methods

2.1. Study Sites

The study was carried out in farms located in the Caquetá Department of Colombia, within a range of 1°19′13.2″ N to 1°44′37.51″ N, and 75°15′40.69″ W to 75°46′10.4″ W. The area is located in the Amazonia Piedmont of the eastern Andean mountain range in a landscape mostly dominated by degraded pastures. Average annual rainfall is 3758 mm and mean temperature 25.8 °C, with 1570 h of sunshine per year (adapted from IDEAM [37]).

In the study region, the landscape predominates with rolling hills of slopes lower than 25%. The soils originated from sedimentary parent material native of the Amazonic mega-basin [38]. The mineralogical fraction is constituted by gibbsite ($Al(OH)_3$), kaolinite ($Al_2Si_2O_5(OH)_4$), mica and goethite [α-Fe$_3$$^+$O(OH)] [39]. The soils present low natural fertility, with textures ranging from silty clay to sandy clay loam, low base saturation, extreme acidity, and exchangeable aluminum saturation at toxic levels for most field crops [38]. Some soil characteristics of studied plots are provided in Table 1.

Ten paired areas (from 0.22 to 3.5 ha) with grass-alone (GA) and grass-legume (GL) pastures in adjacent plots were identified in six farms. Informal interviews with the farmers indicated that pastures were aged between 16 and 32 years, and were sown using tillage with a disc harrow and applying between 0.2 to 1.0 Mg ha^{-1} of $CaMg(CO_3)_2$ in five of the farms, but no-tillage or liming was used for establishing pastures in the sixth one (E1-2, Supplementary Table S1). Five of the six farmers have repeated liming since the establishment of the pastures, with intervals of several years (e.g., the plots F1-2 received lime six years before sampling). None of the farms received maintenance fertilizers or renovated pastures by re-sowing them. The grazing management of the pastures is under rotation, usually between one to three days of grazing and between 27 to 45 days of rest to permit recovery and growth of the pasture, with a dual-purpose cattle system for milk production in five farms and beef production in one farm. The establishment and management of pastures differ between farms rather than pasture types. However, sometimes more productive animals graze on GL than on GA (e.g., non-lactating cows grazing in GA), and the grazing duration gets adjusted according to forage availability. Introduced grasses evaluated in the farms were *Urochloa humidicola* cv. Tully (CIAT 679), *U. brizantha* cv. Toledo (CIAT 26110), and *U. decumbens* cv. Basilisk (CIAT 606). The associated forage legumes were either *Arachis pintoi* cv. Mani Forrajero (CIAT 17434) or *Pueraria phaseoloides* cv. Kudzu (CIAT 9900). Detailed information about farms management and establishment of pastures, and species found per farm is provided in Supplementary Tables S1 and S2.

Table 1. Soil chemical characteristics (0–10 cm soil depth) of grass-alone and grass-legume pastures sampled on farms in the Caquetá Department of Colombia. Each value represents the mean of ten plots for grass-alone, and of eight plots for grass-legume pastures.

Pasture Type	pH [a]	Total N (mg g Dry Soil^{-1}) [b]	NH_4^+ (mg kg Dry Soil^{-1}) [c]	NO_3^- (mg kg Dry Soil^{-1}) [c]	Total C (mg g Dry Soil^{-1}) [b]	Bray-II P (mg kg Dry Soil^{-1}) [d]	C:N	δ^{15}N (‰) [b]	δ^{13}C (‰) [b]
Grass-alone	4.8 ± 0.3	2.9 ± 0.5	4.1 ± 1.0	3.1 ± 2.1	32.4 ± 5.6	1.27 ± 0.5	11.2 ± 0.7	5.9 ± 0.7	−20.6 ± 1.6
Grass-legume	4.8 ± 0.1	2.6 ± 0.7	3.9 ± 2.3	3.4 ± 4.5	28.2 ± 7.4	1.09 ± 0.6	10.6 ± 0.8	6.1 ± 1.2	−21.0 ± 1.6

[a] Measured in deionized water. [b] Total N, C, δ^{15}N, and δ^{13}C determined by dry combustion using an NCS elemental analyzer coupled to an Isotope Ratio Mass spectrometer (Vario PYRO cube, Elementar, Germany and IsoPrime100 IRMS, Isoprime, United Kingdom) (precision ± 0.2‰), further details are provided in Section 2.3. [c] Mineral N extraction in 1M KCl and quantification of NH_4^+ and NO_3^- following Borrero et al. [40]. [d] Bray-II extractable P [40,41].

2.2. Plant and Soil Sampling

In each pasture, at the end of May 2019, a 25 m^2 plot was fenced to impede animal grazing and deposition of excreta for 45 days before sampling. In mid-July 2019, one sampling circle of 5 m radius was delimited per plot. Topsoil samples (0–10 cm soil depth) were collected using an Eijkelkamp Edelman soil auger in the center of the circle and in other six points that were equally distant to each other in the periphery. Soil subsamples were then mixed and a composite sample per plot was air-dried for 48 h and passed through a 2 mm sieve. A PVC frame of 1 m^2 was placed randomly inside each sampling circle. Shoot biomass in the frames was cut to ground level, and harvested after 45 days of regrowth. This relatively long regrowth period was required, because the period fell into the rainy season, with 356, 473 and 611 mm of precipitation during May, June, and July 2019, respectively [42]. Slow rates of regrowth resulted from both, high levels of precipitation and cloudiness during the day, which may have resulted in lower level of photosynthetically active radiation [43], and it may have been partially influenced by the low cutting level (<5.0 cm) used to homogenize the pasture height before harvest. The harvested biomass was split into four botanical fractions: principal grasses (*Urochloa* spp.), secondary grasses (native/naturalized e.g., *Homolepis aturensis* and/or *Paspalum* spp.), legumes, and forbs. A plant litter composite sample was also collected. Plant litter was defined as dead plant parts lying on the ground including dead and completely dry grass leaves still attached to the shoot and senescent legume leaves. Plant samples were oven-dried at 60 °C for 72 h and their dry matter (DM) weight was determined. A subsample of each fraction was ground using a cutting mill (RETSCH, model SM 100) and pulverized using a home-made ball mill with a SIEMENS engine.

At the beginning of the study, we identified 10 paired GA and GL pasture plots. Nevertheless, possibly due to seasonal changes, at the time of the sampling two GL plots showed legume proportions that were lower than 3% and these two plots were not included to the total number of GL plots. Thus, the final results reported were based on 10 GA and 8 GL plots with an average legume proportion of 35% (10–60%) with respect to the total green biomass DM of the plot.

2.3. Chemical and Isotopic Analysis of Plants and Soil

Plant and soil samples were analyzed for total N and C concentration, ^{15}N/^{14}N and ^{13}C/^{12}C isotopic ratios by dry combustion using an NCS elemental analyzer coupled to an Isotope Ratio Mass spectrometer (Vario PYRO cube, Elementar, Germany and IsoPrime100 IRMS, Isoprime, United Kingdom) at ETH Zurich, Eschikon, Switzerland. The natural ^{15}N abundance values are expressed as δ^{15}N, i.e., per mil (‰) ^{15}N excess or depletion over the ^{15}N/^{14}N ratio of the air (R_{air} = 367.6 × 10^{-5}) [44]:

$$\delta^{15}\text{N}\ (‰) = \frac{^{15}\text{N}/^{14}\text{N ratio}_{\text{sample}} - {}^{15}\text{N}/^{14}\text{N ratio}_{\text{air}}}{^{15}\text{N}/^{14}\text{N ratio}_{\text{air}}} \times 1000, \tag{1}$$

The δ^{13}C is accordingly expressed as ^{13}C excess or depletion over the ^{13}C/^{12}C ratio of the international Vienna Pee Dee Belemnite (VPDB) standard (R_{VPDB} = 1123.75 × 10^{-5}). For calibration, we used two international standards, IAEA-N-1 (δ^{15}N = +0.4‰) and IAEA-N-2 (δ^{15}N = +20.3‰), peptone (δ^{15}N = +6.7‰) and glycine (δ^{15}N = +12.2‰) for nitrogen, and peptone (δ^{13}C = −15.7‰) and glycine (δ^{13}C = −33.25‰) for carbon. Correction for instrumental drift was done by repeated measurement of a sulfanilamide internal standard (δ^{13}C = −28‰, δ^{15}N = −0.8‰). Repeated measurement of the sulfanilamide standard gave an analytical precision of 0.3‰ for δ^{15}N, and 0.2‰ for δ^{13}C. Calibrated pea grain was repeatedly measured as an internal quality check (δ^{13}C = −24.7‰, δ^{15}N = +2.4‰).

The P concentration in the plant tissue was determined after digestion of 0.2 g of pulverized leaf tissue with 2 mL of distilled water and 2 mL of concentrated HNO$_3$ using a high-pressure single reaction chamber microwave system (turboWave, MWS microwave systems) [45]. The P concentration in the extracts was determined colorimetrically at 610 nm using the malachite green method [46].

Nutrient (N and P) uptake per botanical fraction was calculated by multiplying their nutrient concentration by the biomass production per m^2. Total nutrient uptake was determined by summing the nutrient uptake of each botanical fraction except plant litter in 1 m^2.

The weighted δ^{15}N of the swards (on a m^2 basis) was determined by applying the formula:

$$\text{Weighted } \delta^{15}\text{N (\textperthousand)} = \frac{[(\delta^{15}\text{N}_{Pg} \times \text{N uptake}_{Pg}) + (\delta^{15}\text{N}_{Sg} \times \text{N uptake}_{Sg}) + (\delta^{15}\text{N}_{Leg} \times \text{N uptake}_{Leg}) + (\delta^{15}\text{N}_{Forbs} \times \text{N uptake}_{Forbs})]}{\text{Total N uptake of the plot}}, \tag{2}$$

where Pg: principal grass, Sg: secondary grass, Leg: legumes, N uptake: N uptake in respective botanical fraction, in g m^{-2}. The weighed δ^{13}C (‰) was calculated accordingly.

The weighted nutrient concentration of N and P of the total biomass per plot:

$$\text{Weighted nutrient concentration } (\text{mg g DM}^{-1}) = \frac{[(\text{NC}_{Pg} \times \text{biomass}_{Pg}) + (\text{NC}_{Sg} \times \text{biomass}_{Sg}) + (\text{NC}_{Leg} \times \text{biomass}_{Leg}) + (\text{NC}_{Forbs} \times \text{biomass}_{Forbs})]}{\text{Total biomass of the plot}} \tag{3}$$

where NC is the nutrient concentrations (N or P, in mg g^{-1}).

2.4. Legume N Derived From the Atmosphere

The proportion of N derived from the atmosphere (%Ndfa, i.e., fixed N) was determined by the ^{15}N natural abundance method [44] and applying the following formula:

$$\text{Ndfa (\%)} = \frac{\delta^{15}\text{N}_{ref} - \delta^{15}\text{N}_{leg}}{\delta^{15}\text{N}_{ref} - B} \times 100, \tag{4}$$

where δ^{15}Nref: δ^{15}N signature of the non-fixing reference plant shoots, δ^{15}Nleg: δ^{15}N signature of the legumes shoots, B: is the δ^{15}N of *Arachis pintoi* or *Pueraria phaseoloides* shoots relying on atmospheric N$_2$ as a sole source of N and it accounts for any internal isotopic fractionation of legume plants [47].

For reference plant, we used the forbs growing in the sampling area, i.e., in association with the legume. The δ^{15}N of non-N$_2$-fixing reference plant was assumed to be identical to the δ^{15}N of soil N taken up by the legume [48]. B values used in our study were obtained from previous reports for the legume species studied or the most closely related species of the same genus. For *Pueraria phaseoloides* the B value used was −1.22 [49]. For *Arachis pintoi* we used −0.88, the mean of three values reported for *Arachis hypogea* [49–51].

The amount of N fixed was calculated for each GL plot by multiplying the N uptake in shoot DM of legumes with the respective %Ndfa.

2.5. Data Analysis

Statistical analyses and figures were performed using R v3.4.4. Significant differences between botanical fractions were assessed through a linear mixed-effects model treating pasture type and botanical fraction as fixed factors, and farm as random effect using the packages 'lme4' v1.1-23 and 'nlme' v3.1-147 in R. Multiple comparisons between botanical fractions were evaluated with TukeyHSD tests using 'emmeans' v1.4.8. To evaluate total differences between pasture types, weighted nutrient concentrations and isotopic signatures were calculated per farm, and a second model was built treating pasture type as fixed factor and farm as random effect. Analysis of δ^{15}N of grass and legume species followed the same model structure as for botanical fractions, considering species and pasture type as fixed factors, and farm as random effect. Differences in %Ndfa between legume species were not statistically tested due to very high differences in sample size ($n = 6$ for *A. pintoi*, and $n = 2$ for *P. phaseoloides*). The correlation between δ^{15}N of grasses and forbs, and the δ^{15}N of grasses with grass biomass and N concentration were tested with the Pearson's correlation coefficient (r). Figures were constructed using 'ggplot2' v2.2.1.

3. Results

3.1. Dry Matter Productivity and Nutrient Uptake

Grass-legume pastures produced more plant biomass and had greater nutrient (N and P) uptake than GA pastures. Excluding the plant litter fraction, the extent of increase in GL compared to GA swards was up to 74% for shoot DM production (g DM m^{-2}: 62 in GA vs 108 in GL), while it was more than two-fold higher for N uptake (g N m^{-2}: 0.8 in GA vs 2.2 in GL) and P uptake (g P m^{-2}: 0.07 in GA vs. 0.14 in GL) (Figure 1). The proportion of biomass of forbs in the total plant biomass of the pastures was lower in the GL (3% of total DM) than in the GA pastures (16% of total DM).

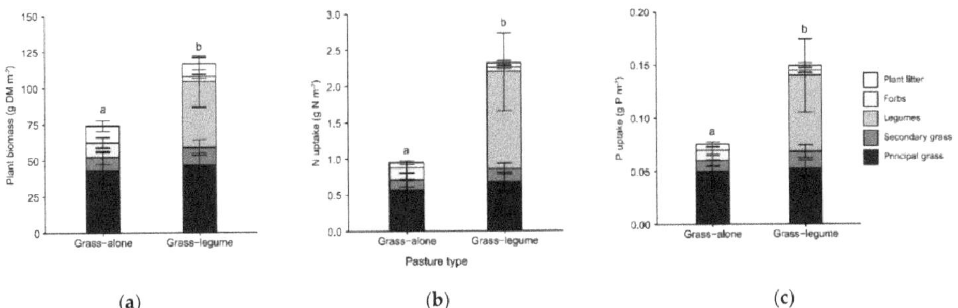

Figure 1. (**a**) Shoot plant biomass production (**b**) N uptake and (**c**) P uptake in grass alone and grass-legume pastures for each plant fraction; $n = 10$ for grass alone and 8 for grass-legume pastures. DM = dry matter. Different letters denote statistical differences for the total plant biomass production, N and P uptake of the sward per pasture type ($\alpha = 0.05$).

As expected, N concentrations were significantly higher in legume than grass shoots (Table 2). In contrast, P concentrations did not differ significantly between grasses and legumes. Weighted N concentration of GL plant biomass was 18% higher than that of the biomass in GA plots. No difference was observed in weighted plant biomass P concentration between the two pasture types. Therefore, the higher P uptake observed in GL than GA pastures resulted from greater DM production.

Table 2. Nutrient concentrations (N, P, C) and isotopic signatures of N and C in two different pasture types for each botanical fraction. Values are mean ± standard deviation, with $n = 10$ for grass alone, and $n = 8$ for grass-legume pastures. DM = dry matter, ns = not significant.

Pasture Type	Botanical Fraction	N Concentration (g N kg DM^{-1})	P Concentration (g P kg DM^{-1})	C Concentration (g C kg DM^{-1})	C:N	δ^{15}N (‰)	δ^{13}C (‰)
Grass alone	Forbs	18.8 ± 4.1 c	1.3 ± 0.4 b	411.1 ± 5.4 a	22.7 ± 4.7 a	5.3 ± 1.1 d	−25.0 ± 5.1 b
	Principal grass	14.9 ± 3.5 b	1.1 ± 0.3 b	424.0 ± 5.4 b	30.0 ± 7.6 a	4.5 ± 3.1 c	−13.3 ± 0.5 d
	Secondary grass	15.3 ± 5.3 bc	1.4 ± 0.3 b	420.5 ± 8.1 ab	30.5 ± 11.4 a	2.5 ± 3.0 c	−20.9 ± 6.2 c
	Legumes	22.1 ± 4.8 d	1.2 ± 0.1 b	447.9 ± 9.0 c	20.7 ± 4.9 a	−0.6 ± 0.6 a	−31.0 ± 0.0 a
	Plant litter	6.8 ± 1.8 a	0.4 ± 0.1 a	424.8 ± 11.7 b	67.1 ± 20.1 b	1.9 ± 2.3 b	−16.9 ± 3.8 c
	Total *	17.4 ± 3.9 A	1.2 ± 0.3 A	426.2 ± 11.5 A	25.8 ± 7.1 A	4.6 ± 2.9 B	−17.1 ± 3.8 A
Grass legume	Forbs	18.4 ± 3.1 c	1.4 ± 0.4 b	417.4 ± 7.5 ab	23.1 ± 3.7 a	5.7 ± 2.5 d	−25.4 ± 5.3 b
	Principal grass	14.8 ± 3.5 b	1.3 ± 0.5 b	426.9 ± 5.0 b	30.1 ± 6.8 a	3.8 ± 2.9 c	−12.7 ± 0.4 d
	Secondary grass	15.4 ± 2.1 bc	1.4 ± 0.3 b	419.8 ± 5.0 ab	27.6 ± 3.7 a	3.6 ± 3.2 c	−18.0 ± 3.8 c
	Legumes	27.8 ± 3.3 d	1.3 ± 0.3 b	422.1 ± 9.2 ab	15.3 ± 1.8 a	0.4 ± 1.0 a	−29.6 ± 0.3 a
	Plant litter	7.9 ± 1.3 a	0.5 ± 0.2 a	417.1 ± 8.6 ab	53.6 ± 7.3 b	1.1 ± 2.1 b	−16.0 ± 0.9 c
	Total *	20.5 ± 3.2 B	1.2 ± 0.2 A	404.9 ± 46.2 A	20.0 ± 3.3 A	2.1 ± 1.7 A	−19.9 ± 3.5 A
Source of variation **	Pasture type	ns	ns	$p < 0.05$	ns	ns	ns
	Botanical fraction	$p < 0.001$	$p < 0.001$	$p < 0.001$	$p < 0.001$	$p < 0.001$	$p < 0.001$
	Pasture type x botanical fraction	ns	ns	$p < 0.001$	ns	ns	ns
	Farm (random)	$p < 0.001$	$p < 0.001$	$p < 0.05$	ns	$p < 0.001$	$p < 0.001$

* Calculated values of weighted nutrient concentration or isotopic signature of each botanical fraction (forbs, grasses and legumes only) by their nutrient uptake as in Equations (2) and (3). ** Sources of variation apply only for comparison of botanical fractions. For each variable, different lowercase letters indicate statistical differences of botanical fractions within and between pasture types. Uppercase letters indicate statistical differences for the total (weighted) concentrations between pasture types according to the Tukey HSD test ($\alpha = 0.05$).

3.2. Legume-N derived from the atmosphere

The weighted $\delta^{15}N$ signature of the combined plant biomass of GA pastures was 4.6‰, while it was 2.1‰ for GL pastures (Table 2). The $\delta^{15}N$ signature of forbs was similar to that of soil N, whereas that of principal grasses was on average by 1.4‰ (GA) and 2.3‰ (GL) less enriched than soil N (Tables 1 and 2, Figure 2). Therefore, forbs were considered as more appropriate reference plants to determine Ndfa than grasses.

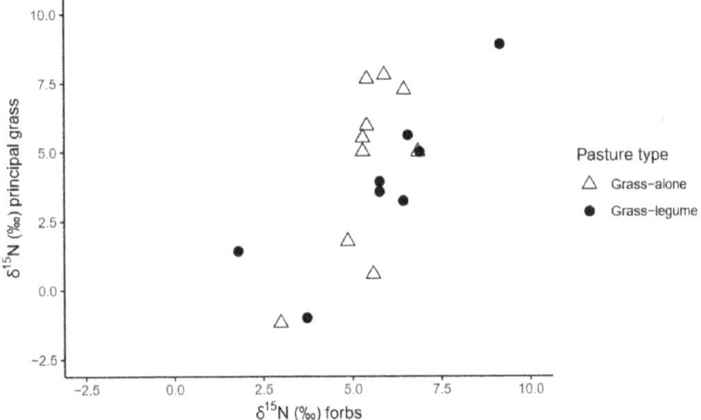

Figure 2. Relationship between shoot $\delta^{15}N$ signature of forbs and principal grasses sampled in grass-alone (open triangles, $n = 10$) and grass-legume (full circles, $n = 8$) pastures. Pearson's correlation coefficient (r) = 0.76, $p < 0.01$.

Arachis pintoi was the legume species found in six out of eight farms, and *P. phaseoloides* occurred in two farms. Average Ndfa derived for GL pastures using Equation (4) ranged from 60% to 99% (average value of 80%, Table 3). *A. pintoi* showed on average 16% higher %Ndfa than *P. phaseoloides*. The amount of N fixed in the biomass of legumes ranged from 0.15 to 3.7 g N m^{-2}.

Table 3. Grass and legume species, average $\delta^{15}N$ signature of shoots and %Ndfa ± standard deviation of the mean observed per species. Ndfa = Nitrogen derived from the atmosphere, ns = not significant.

Pasture Type	Species	$\delta^{15}N$ (‰)	Ndfa (%)
Grass alone	*U. brizantha*	4.7 ± 4.2 ab	-
	U. decumbens	5.8 ± 1.1 b	-
	U. humidicola	3.2 ± 4.2 ab	-
Grass legume	*U. brizantha*	4.9 ± 2.7 ab	-
	U. decumbens	5.0 ± 0.0 ab	-
	U. humidicola	2.0 ± 3.3 ab	-
	A. pintoi	0.4 ± 0.9 a	83.2 ± 14.0
	P. phaseoloides	1.3 ± 0.6 ab	67.5 ± 9.2
Source of variation	Pasture type	ns	-
	Species	$p < 0.05$	-
	Pasture type x species	ns	-
	Farm (random)	ns	-

Different letters indicate statistical differences according to the TukeyHSD test ($\alpha = 0.05$).

3.3. $\delta^{15}N$ and $\delta^{13}C$ Isotopic Signature of Pasture Components

Urochloa humidicola was the principal grass in four out of ten GA plots and three out of eight GL plots, while it was *U. decumbens* in four GA plots and one GL, and *U. brizantha* in two GA and four GL plots (Supplementary Table S2). The $\delta^{15}N$ signature of principal grasses varied widely among plots of the same type of pasture. In three out of ten plots, the GA-principal grass showed δ^{15}N lower than 2‰, and in two of them (A2 and F1), even lower than the corresponding GL principal grass signature (Supplementary Table S2). The average δ^{15}N signature of shoot tissue of *U. humidicola* tended to be lower than that of *U. decumbens* and *U. brizantha*, however, this difference was not statistically significant (Table 3).

The δ^{15}N of the principal grass was negatively related to the DM production of the principal grass (r = −0.5, p < 0.05. Figure 3a), and positively related to the N concentration of the principal grass shoot tissue (r = 0.5, p < 0.05. Figure 3b). However, this latter correlation was stronger in the GA pastures (r = 0.68, p < 0.05) than in GL (r = 0.26, non-significant). At N concentrations higher than 14 mg g^{-1} in the shoot tissue, the δ^{15}N of grasses growing in GL pastures was lower than that of GA pastures by at least 2‰ (p < 0.01).

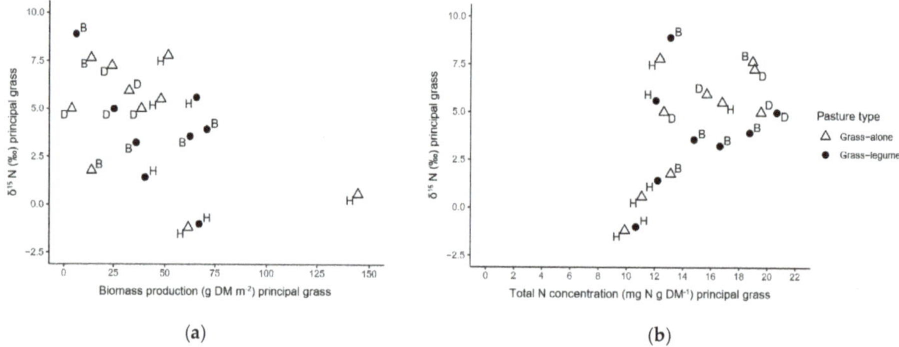

Figure 3. (**a**) Relationship between biomass production and δ^{15}N signature of principal grass shoot, r = −0.5, p < 0.05. (**b**) Relationship between total N concentration and δ^{15}N signature of principal grass, r = 0.5, p < 0.05. Open triangles represent grass-alone pastures, n = 10. Full circles represent grass-legume pastures, n = 8. H = *Urochloa humidicola*, D = *U. decumbens*, B = *U. brizantha*.

The δ^{15}N values of the principal grass and the forbs were closely related (r = 0.76, p < 0.01. Figure 2). The δ^{15}N value of the principal grass was lower than that of the forbs, on average by 0.8‰ for GA pastures, and by 1.9‰ in GL pastures (Table 2, Figure 2).

The δ^{13}C signature of grasses was significantly higher (less negative) than that of forbs and legumes (Table 2). The type of pasture had no significant effect on δ^{13}C of the botanical fractions and soil organic matter (Tables 1 and 2, respectively).

4. Discussion

4.1. Legumes Improve Pasture Productivity and Nutrient Uptake

The average biomass production of 62 g m^{-2} in GA pastures was in the range obtained by Fonte et al. [26] on farms in the same study region (average values of 101 g m^{-2} for *Urochloa* spp. GA pastures which farmers had characterized as productive, and 47 g m^{-2} for degraded pastures). Moreover, during the rainy season, as in our study and Fonte et al. [26], Gomez et al. [43] harvested from fertilized *Urochloa* spp. GA pastures on average 250 g m^{-2} after 42 days of regrowth of the pasture. Pasture biomass production was reduced by a factor of two to three during the wettest months compared to drier months [43], and June 2019 had a very high amount of precipitation. Lower biomass production during the wettest months was probably due to low soil oxygen levels caused by waterlogging and/or lower rates of net photosynthesis associated with a lower level of photosynthetically active radiation, which may have been driven by higher cloudiness [52].

The pastures containing legumes (i.e., GL) had a significantly higher biomass production (Figure 1a). This is in agreement with earlier results obtained from on-station experiments. For instance, the DM production of GL pastures composed by *U. decumbens* and *Calopogonium mucunoides* was 1.25 times greater than that of GA pastures containing *U. decumbens* alone in an experiment near Campo Grande, Brazil [53], and DM production was doubled in experiments with *U. decumbens* and *A. pintoi* association, as compared to *U. decumbens* alone in Carimagua, Colombia [17].

Because *Urochloa* grasses obtain atmospheric N_2 via non-symbiotic N_2 fixation [32,33,54], we used the $\delta^{15}N$ of the forbs as an indicator of the $\delta^{15}N$ of available soil N. The strong depletion of $\delta^{15}N$ in legumes compared to the forbs and grasses indicated that legumes largely relied on atmospheric N (Tables 2 and 3) [44]. The average value of 35% legume DM in the total pasture biomass is within the range of 35–45% proposed by Thomas [20] required to maintain a balanced N cycle of tropical GL pastures with a herbage utilization of 50–70% by grazing animals. The proportion of N derived from the atmosphere (%Ndfa) observed with the legumes in this study was about 80% and this value is at the high end of the range reported for tropical legumes [19,55]. This is remarkable because available P concentrations of less than 2 mg kg soil^{-1} are considered as plant growth-limiting for both grasses and legumes [56,57] and could, therefore, limit N_2 fixation of legumes [19].

The amount of N fixed in the shoots of legumes observed in this study was 0.15 to 3.7 g N m^{-2} for 45 days of regrowth. Thus, the average increase in N uptake of 1.4 g m^{-2} in GL pastures than GA pastures was largely due to N_2 fixing ability by the legume (Figure 1b). Because on average around 80% was derived from N_2 fixation, some legume N transfer to grasses might still explain the overall greater N uptake in GL than GA pastures. In temperate GL meadows, about 50% of the grass N was derived from the associated legumes [29]. Trannin et al. [30] suggested that the mineralization of root residues from *Stylosanthes guianensis* was the major source of legume-N transferred to the associated *Urochloa decumbens*. Moreover, greater N deposition through litter has been reported for GL pastures than GA pastures [58].

The P uptake was doubled in GL compared to the GA pasture (Figure 1c), at similarly low available P status of the soil (Table 1). It was reported that legumes such as *A. pintoi* acquire more P from less available P pools from acid soil with its smaller root system than *Urochloa* grass [59]. Thus, the increase of 72 mg P m^{-2} acquired by GL pasture can be attributed to the superior performance of legume towards improved P cycling in the system through soil P pools [60]. According to the farmers' information, the GL pastures are grazed by more productive animals than the GA pastures, and more P may hence be exported via animal products from plots containing legumes, as suggested by Oberson et al. [61] for *Urochloa*-Kudzu pastures. Thus, legumes may take up more soil P and stimulate biological P cycling (through plant litter, animal excreta and microbial turnover) to keep it in available P forms. Although this could increase the risk of soil P mining [62], the strategic application of small amounts of P fertilizer (10 kg ha^{-1} every two years as maintenance fertilizer) may overcome the risk for soil P mining in grazed pastures [63].

4.2. What N Sources are Exploited by Grasses in Each Pasture Type

The $\delta^{15}N$ of the principal grasses, all of which were *Urochloa* spp. grasses was higher than that of the legumes except in two cases (Table 3, Supplementary Table S2). At the same time, their $\delta^{15}N$ was lower than that of the associated forbs (Table 2, Figure 2). This observation indicates that the grasses, on average, were benefiting less from atmospheric N_2 than the legumes, irrespective of the underlying process.

The N concentrations of *Urochloa* spp. grasses observed in our study were higher than previous reports [64], although it was not sufficient to sustain higher plant growth, as indicated by the lower value of biomass production. At N concentrations higher than 14 mg g^{-1}, the $\delta^{15}N$ of the principal grasses of GL was lower than that of GA pastures by at least 2‰. This suggests that legume N transfer was a process involved in the provision of atmospheric N to grasses. Atmospheric N_2 fixation seems to make a significant contribution resulting in low $\delta^{15}N$ values of the *Urochloa* grasses in both GA and GL pastures (Table 3). Still, the contribution from non-symbiotic N_2 fixation in GA may not be adequate to sustain grass growth without the supply of N from the soil through mineralization. This finding is consistent with earlier reports that suggested no more than 20–40% of N in *Urochloa* grasses was derived from the atmosphere via non-symbiotic N_2 fixation [32,33].

C_4 grasses typically have $\delta^{13}C$ higher than $-20‰$, whereas C_3 legumes usually have $\delta^{13}C$ lower than $-20‰$ due to differences in C isotopic fractionation during CO_2 assimilation [65]. In our study, the $\delta^{13}C$ and $\delta^{15}N$ of soil was not statistically different between pasture types, around $-20‰$ and $6‰$, respectively (Tables 1 and 2). While the former forest C_3 vegetation still affects the isotopic composition of soil organic matter C [66], the contribution of legume residues seems to not have been enough to enrich the total soil N pool.

Although the grasses with the lowest $\delta^{15}N$ and N concentration in shoot tissue were mostly *U. humidicola* (Figure 3b), no clear pattern of distribution was observed among *Urochloa* species, either for plant biomass production or N concentration (Figure 3a, b). We consider that the distribution of grass species observed in our study is representative of the pastures in the region. However, to draw valid conclusions on N uptake and utilization at the species level, a more balanced design with an equal number of observations per species will be needed in future research.

Low ^{15}N natural abundance in the shoot tissue of *U. humidicola* has been interpreted as an indicator of high capacity of biological nitrification inhibition (BNI) in that grass [31]. Indeed, in our study, *U. humidicola* showed the lowest $\delta^{15}N$ values, both in GA and GL pastures, but this grass was found to obtain a relatively significant proportion of N through non-symbiotic N_2 fixation [33]. Our results rather suggest that low $\delta^{15}N$ of the grasses is an indicator of low N availability in soil [67], and grasses adapted to N depleted soils can cope with either through BNI ability [31] and/or with non-symbiotic N_2 fixation [33].

5. Conclusions

In farmers' long-term tropical pastures established in the forest margins of Colombia, legumes associated with grasses (GL pastures) resulted in greater pasture biomass production than grass-alone (GA) pastures. Legumes derived on average 80% of their N from symbiotic N_2 fixation, despite low fertility acid soils with low plant-available P content. Legumes significantly increased both N and P uptake by the pasture biomass. The greater N uptake by legumes could be assigned mostly to N fixed from the atmosphere. The $\delta^{15}N$ signatures of grasses in GA vs. GL pastures suggested that sources of grass N are affected by legumes integrated in the pasture. While lower $\delta^{15}N$ values of grasses growing in GL than GA pastures suggest that grasses could obtain fixed N via legume N transfer, exceptionally low $\delta^{15}N$ values of grasses in GA pastures indicate significant potential for N input via non-symbiotic N_2 fixation from the atmosphere. Overall, this study indicates that *Urochloa* grasses are capable to swap N sources when these grasses are grown in association with legumes. The role of different soil-plant processes such as BNI or N_2 fixation from the atmosphere need to be further studied under field and also controlled conditions. This missing knowledge is critical to define the sources of N for grass growth, either in GA or GL pastures in the tropics.

Supplementary Materials: The following are available online at http://www.mdpi.com/1424-2818/12/11/419/s1, Table S1: Establishment and management parameters of ten grass-alone (GA) and eight grass-legume (GL) paired pastures in six farms in the Caquetá Department of Colombia, Table S2: Grass and legume species, $\delta^{15}N$ signature of shoots, and %Ndfa observed per species per farm.

Author Contributions: Conceptualization: A.O. and J.V.; methodology: A.O., J.V., G.R.; investigation: D.M.V., J.V., G.R., K.O.; data curation: D.M.V.; validation: A.O., J.V., J.A., I.M.R.; writing—original draft preparation: D.M.V. and A.O.; writing—review and editing: D.M.V., J.V., G.R., J.A., I.M.R., A.O.; project administration: A.O., J.V.; funding acquisition: A.O., J.V., J.A.; All authors have read and agreed to the published version of the manuscript.

Funding: The authors acknowledge the financial support of the Leading House for the Latin American Region, University of St. Gallen, Switzerland. D.V. and J.A. acknowledge the UK Research and Innovation (UKRI) Global Challenges Research Fund (GCRF) *GROW Colombia* grant via the UK's Biotechnology and Biological Sciences Research Council (BB/P028098/1). Additional funding for this study was received from the CGIAR Research Programs on Livestock; and Climate Change, Agriculture and Food Security (CCAFS).

Acknowledgments: We warmly acknowledge the great cooperation of farmers in the region. We thank Federica Tamburini (ETH Zurich) for mass spectrometry measurements.

Conflicts of Interest: The authors declare no conflict of interest. The funders had no role in the design of the study; in the collection, analyses, or interpretation of data; in the writing of the manuscript, or in the decision to publish the results.

References

1. Hansen, M.C.; Potapov, P.V.; Moore, R.; Hancher, M.; Turubanova, S.A.; Tyukavina, A.; Thau, D.; Stehman, S.V.; Goetz, S.J.; Loveland, T.R.; et al. High-resolution global maps of 21st-century forest cover change. *Science* **2013**, *342*, 850–853. [CrossRef] [PubMed]
2. Asner, G.P.; Townsend, A.R.; Bustamante, M.M.C.; Nardoto, G.B.; Olander, L.P. Pasture degradation in the central Amazon: Linking changes in carbon and nutrient cycling with remote sensing. *Glob. Chang. Biol.* **2004**, *10*, 844–862. [CrossRef]
3. Jank, L.; Barrios, S.C.; Do Valle, C.B.; Simeáo, R.M.; Alves, G.F. The value of improved pastures to Brazilian beef production. *Crop Pasture Sci.* **2014**, *65*, 1132–1137. [CrossRef]
4. Wassenaar, T.; Gerber, P.; Verburg, P.H.; Rosales, M.; Ibrahim, M.; Steinfeld, H. Projecting land use changes in the Neotropics: The geography of pasture expansion into forest. *Glob. Environ. Chang. Hum. Policy Dimens.* **2007**, *17*, 86–104. [CrossRef]
5. Jimenez, J.J.; Lal, R. Mechanisms of C sequestration in soils of Latin America. *Crit. Rev. Plant Sci.* **2006**, *25*, 337–365. [CrossRef]
6. Boddey, R.M.; Macedo, R.; Tarre, R.M.; Ferreira, E.; de Oliveira, O.C.; Rezende, C.D.; Cantarutti, R.B.; Pereira, J.M.; Alves, B.J.R.; Urquiaga, S. Nitrogen cycling in *Brachiaria* pastures: The key to understanding the process of pasture decline. *Agric. Ecosyst. Environ.* **2004**, *103*, 389–403. [CrossRef]
7. De Oliveira, O.C.; de Oliveira, I.P.; Alves, B.J.R.; Urquiaga, S.; Boddey, R.M. Chemical and biological indicators of decline/degradation of *Brachiaria* pastures in the Brazilian Cerrado. *Agric. Ecosyst. Environ.* **2004**, *103*, 289–300. [CrossRef]
8. Nesper, M.; Buenemann, E.K.; Fonte, S.J.; Rao, I.M.; Velasquez, J.E.; Ramirez, B.; Hegglin, D.; Frossard, E.; Oberson, A. Pasture degradation decreases organic P content of tropical soils due to soil structural decline. *Geoderma* **2015**, *257*, 123–133. [CrossRef]
9. Steinfield, H.; Gerber, P.; Wassenaar, T.; Castel, V.; Rosales, M.; de Haan, C. *Livestock's Long Shadow: Environmental Issues and Options*; FAO: Rome, Italy, 2006.
10. FAO. Country Pasture/Forage Resource Profiles Brazil. 2006. Available online: http://www.fao.org/ag/agp/AGPC/doc/Counprof/Brazil/brazil2.htm (accessed on 10 June 2020).
11. Navarrete, D.; Sitch, S.; Aragão, L.E.O.C.; Pedroni, L. Conversion from forests to pastures in the Colombian Amazon leads to contrasting soil carbon dynamics depending on land management practices. *Glob. Chang. Biol.* **2016**, *22*, 3503–3517. [CrossRef]
12. Dias-Filho, M.B.; Davidson, E.A.; de Carvalho, C.J.R. Linking biochemical cycles to cattle pasture management and sustainability in the Amazon Basin. In *The Biogeochemistry of the Amazon Basin*; McClain, M.E., Victoris, R.L., Richery, J.E., Eds.; Oxford University Press: New York, NY, USA, 2001; pp. 84–105.
13. Fisher, M.J.; Braz, S.P.; Dos Santos, R.S.M.; Urquiaga, S.; Alves, B.J.R.; Bodddey, R.M. Another dimension to grazing systems: Soil carbon. *Trop. Grassl.* **2007**, *41*, 65–83.
14. Neill, C.; Piccolo, M.C.; Melillo, J.M.; Steudler, P.A.; Cerri, C.C. Nitrogen dynamics in Amazon forest and pasture soils measured by N-15 pool dilution. *Soil Biol. Biochem.* **1999**, *31*, 567–572. [CrossRef]
15. Miles, J.W.; Do Valle, C.B.; Rao, I.M.; Euclides, V.P.B. Brachiaria grasses. In *Warm Season (C4) Grasses*; ASA, CSSA, SSSA: Madison, WI, USA, 2004; Volume Monograph No. 45, p. 39.
16. Rao, I.M. Adapting tropical forages to low-fertility soils. In Proceedings of the Xix International Grassland Congress, São Pedro, São Paulo, Brazil, 11–21 February 2001; pp. 247–254.
17. Fisher, M.J.; Kerridge, P.C. The agronomy and physiology of *Brachiaria* species. In *Brachiaria: Biology, Agronomy, and Improvement*; Miles, J.W., Maass, B.L., do Valle, C.B., Eds.; Centro International de Agricultura Tropical: Cali, Colombia, 1996; pp. 43–52.
18. Schultze-Kraft, R.; Rao, I.M.; Peters, M.; Clements, R.J.; Bai, C.; Liu, G. Tropical forage legumes for environmental benefits: An overview. *Trop. Grassl. Forrajes Trop.* **2018**, *6*, 1–14. [CrossRef]

19. Cadisch, G.; Sylvester-Bradley, R.; Nösberger, J. ^{15}N-based estimation of nitrogen fixation by eight tropical forage-legumes at two levels of P:K supply. *Field Crops Res.* **1989**, *22*, 181–194. [CrossRef]

20. Thomas, R.J. The role of the legume in the nitrogen cycle of productive and sustainable pastures. *Grass Forage Sci.* **1992**, *47*, 133–142. [CrossRef]

21. Lüscher, A.; Mueller-Harvey, I.; Soussana, J.F.; Rees, R.M.; Peyraud, J.L. Potential of legume-based grassland-livestock systems in Europe: A review. *Grass Forage Sci.* **2014**, *69*, 206–228. [CrossRef]

22. Cipagauta, M.; Velásquez, J.; Pulido, J.I. Producción de leche en tres pasturas del Piedemonte Amazonico del Caquetá, Colombia. *Pasturas Trop.* **1998**, *20*, 2–10.

23. Boddey, R.M.; Casagrande, D.R.; Homem, B.G.C.; Alves, B.J.R. Forage legumes in grass pastures in tropical Brazil and likely impacts on greenhouse gas emissions: A review. *Grass Forage Sci.* **2020**. [CrossRef]

24. Rao, I.; Peters, M.; Castro, A.; Schultze-Kraft, R.; White, D.; Fisher, M.; Miles, J.; Lascano, C.; Blümmel, M.; Bungenstab, D.; et al. LivestockPlus—The sustainable intensification of forage-based agricultural systems to improve livelihoods and ecosystem services in the tropics. *Trop. Grassl. Forrajes Trop.* **2015**, *3*, 59–82. [CrossRef]

25. Gaviria-Uribe, X.; Naranjo-Ramírez, J.F.; Bolívar-Vergara, D.M.; Barahona-Rosales, R. Intake and digestibility of nutrients in zebu steers grazing in intensive silvopastoral system. *Arch. Zootec.* **2015**, *64*, 21–27. [CrossRef]

26. Fonte, S.J.; Nesper, M.; Hegglin, D.; Velásquez, J.E.; Ramirez, B.; Rao, I.M.; Bernasconi, S.M.; Bünemann, E.K.; Frossard, E.; Oberson, A. Pasture degradation impacts soil phosphorus storage via changes to aggregate-associated soil organic matter in highly weathered tropical soils. *Soil Biol. Biochem.* **2014**, *68*, 150–157. [CrossRef]

27. Nakamura, S.; Saliou, P.S.; Takahashi, M.; Ando, Y.; Subbarao, G.V. The contribution of root turnover on biological nitrification inhibition and its impact on the ammonia-oxidizing archaea under *Brachiaria* cultivations. *Agronomy* **2020**, *10*, 1003. [CrossRef]

28. Hammelehle, A.; Oberson, A.; Luscher, A.; Mader, P.; Mayer, J. Above- and belowground nitrogen distribution of a red clover-perennial ryegrass sward along a soil nutrient availability gradient established by organic and conventional cropping systems. *Plant Soil* **2018**, *425*, 507–525. [CrossRef]

29. Oberson, A.; Frossard, E.; Bühlmann, C.; Mayer, J.; Mäder, P.; Lüscher, A. Nitrogen fixation and transfer in grass-clover leys under organic and conventional cropping systems. *Plant Soil* **2013**, *371*, 237–255. [CrossRef]

30. Trannin, W.S.; Urquiaga, S.; Guerra, G.; Ibijbijen, J.; Cadisch, G. Interspecies competition and N transfer in a tropical grass-legume mixture. *Biol. Fertil. Soils* **2000**, *32*, 441–448. [CrossRef]

31. Karwat, H.; Egenolf, K.; Nuñez, J.; Rao, I.; Rasche, F.; Arango, J.; Moreta, D.; Arevalo, A.; Cadisch, G. Low ^{15}N natural abundance in shoot tissue of *Brachiaria humidicola* is an indicator of reduced N losses due to biological nitrification inhibition (BNI). *Front. Microbiol.* **2018**, *9*, 2383. [CrossRef]

32. Boddey, R.; Victoria, R. Estimation of biological nitrogen fixation associated with *Brachiaria* and *Paspalum* grasses using ^{15}N labelled organic matter and fertilizer. In *Nitrogen Fixation with Non-Legumes*; Springer: Dordrecht, The Netherlands, 1986; pp. 265–292.

33. Reis, V.M.; dos Reis, F.B.; Quesada, D.M.; de Oliveira, O.C.A.; Alves, B.J.R.; Urquiaga, S.; Boddey, R.M. Biological nitrogen fixation associated with tropical pasture grasses. *Aust. J. Plant Physiol.* **2001**, *28*, 837–844. [CrossRef]

34. Nunez, J.; Arevalo, A.; Karwat, H.; Egenolf, K.; Miles, J.; Chirinda, N.; Cadisch, G.; Rasche, F.; Rao, I.; Subbarao, G.; et al. Biological nitrification inhibition activity in a soil-grown biparental population of the forage grass, *Brachiaria humidicola*. *Plant Soil* **2018**, *426*, 401–411. [CrossRef]

35. Subbarao, G.V.; Nakahara, K.; Hurtado, M.P.; Ono, H.; Moreta, D.E.; Salcedo, A.F.; Yoshihashi, A.T.; Ishikawa, T.; Ishitani, M.; Ohnishi-Kameyama, M.; et al. Evidence for biological nitrification inhibition in *Brachiaria* pastures. *Proc. Natl. Acad. Sci. USA* **2009**, *106*, 17302–17307. [CrossRef]

36. Nyfeler, D.; Huguenin-Elie, O.; Suter, M.; Frossard, E.; Lüscher, A. Grass-legume mixtures can yield more nitrogen than legume pure stands due to mutual stimulation of nitrogen uptake from symbiotic and non-symbiotic sources. *Agric. Ecosyst. Environ.* **2011**, *140*, 155–163. [CrossRef]

37. Instituto de Hidrología Meteorología y Estudios Ambientales (IDEAM). Atlas Climatológico, Radiación y de Viento. Available online: http://www.ideam.gov.co/web/tiempo-y-clima/ (accessed on 7 August 2020).

38. Instituto Geográfico Agustín Codazzi (IGAC). *Estudio General de Suelos y Zonificación de Tierras: Departamento de Caquetá. Escala 1:100.000*; Instituto Geográfico Agustín Codazzi (IGAC): Bogotá, Colombia, 2014.

39. Rosas-Patiño, G.; Puentes-Páramo, Y.J.; Menjivar-Flores, J.C. Liming effect on macronutrient intake for cacao (*Theobroma cacao* L.) in the Colombian Amazon. *Cienc. Tecnol. Agropecu.* **2019**, *20*, 5–28.

40. Borrero Tamayo, G.; Jiménez, J.; Ricaurte Oyola, J.J.; Rivera, M.; Polanía Perdomo, J.A.; Núñez, J.; Barbosa, N.; Arango, J.; Cardoso, J.; Rao, I.M. Manual de protocolos. In *Nutrición y Fisiología de Plantas-Forrajes y Fríjol*; Centro Internacional de Agricultura Tropical (CIAT): Cali, Colombia, 2017.

41. Murphy, J.; Riley, J.P. A modified single solution method for the determination of phosphate in natural waters. *Anal. Chim. Acta* **1962**, *27*, 31–36. [CrossRef]

42. Instituto de Hidrología Meteorología y Estudios Ambientales (IDEAM). Geoportal de datos hidrometeorológicos DHIME. Available online: http://www.dhime.ideam.gov.co/ (accessed on 13 August 2020).

43. Gómez, M.; Velásquez, J.; Miles, J.; Rayo, F. Adaptation of Brachiaria to the Amazonian piedmont in Colombia. *Pasturas Trop.* **2000**, *22*, 19–25.

44. Shearer, G.; Kohl, D.H. N2 fixation in field settings: Estimations based on natural abundance. *Aust. J. Plant Physiol.* **1986**, *13*, 699–744. [CrossRef]

45. Wiggenhauser, M.; Bigalke, M.; Imseng, M.; Keller, A.; Archer, C.; Wilcke, W.; Frossard, E. Zinc isotope fractionation during grain filling of wheat and a comparison of zinc and cadmium isotope ratios in identical soil–plant systems. *New Phytol.* **2018**, *219*, 195–205. [CrossRef]

46. Ohno, T.; Zibilske, L.M. Determination of low concentrations of phosphorus in soil extracts using malachite green. *Soil Sci. Soc. Am. J.* **1991**, *55*, 892–895. [CrossRef]

47. Unkovich, M.J.; Pate, J.S.; Sanford, P.; Armstrong, E.L. Potential precision of the d^{15}N natural abundance method in the field estimates of nitrogen fixation by crop and pasture legumes in south-west Australia. *Aust. J. Agric. Res.* **1994**, *45*, 119–132. [CrossRef]

48. Unkovich, M.; Herridge, D.; Peoples, M.; Cadisch, G.; Boddey, R.; Giller, K.; Alves, B.; Chalk, P.M. *Measuring Plant-Associated Nitrogen Fixation in Agricultural Systems*; ACIAR: Canberra, Australia, 2008; Volume 136.

49. Peoples, M.; Bell, M.; Bushby, H. Effect of rotation and inoculation with *Bradyrhizobium* on nitrogen fixation and yield of peanut (*Arachis hypogaea* L., cv. Virginia Bunch). *Aust. J. Agric. Res.* **1992**, *43*, 595–607. [CrossRef]

50. Cadisch, G.; Hairiah, K.; Giller, K.E. Applicability of the natural N-15 abundance technique to measure N-2 fixation in *Arachis hypogaea* grown on an Ultisol. *Neth. J. Agric. Sci.* **2000**, *48*, 31–45.

51. Okito, A.; Alves, B.; Urquiaga, S.; Boddey, R.M. Isotopic fractionation during N-2 fixation by four tropical legumes. *Soil Biol. Biochem.* **2004**, *36*, 1179–1190. [CrossRef]

52. Rao, I.M. Advances in improving adaptation of common bean and *Brachiaria* forage grasses to abiotic stress in the tropics. In *Handbook of Plant and Crop Physiology*; Pessarakli, M., Ed.; CRC Press: Boca Raton, FL, USA, 2014; pp. 847–889.

53. Seiffert, N.; Zimmer, A. Contribución de Calopogonium mucunoides al contenido de nitrógeno en pasturas de *Brachiaria decumbens*. *Pasturas Trop.* **1988**, *10*, 8–13.

54. Ribeiro, N.V.d.S.; Vidal, M.S.; Barrios, S.C.L.; Baldani, V.L.D.; Baldani, J.I. Genetic diversity and growth promoting characteristics of diazotrophic bacteria isolated from 20 genotypes of *Brachiaria* spp. *Plant Soil* **2020**, *451*, 187–205. [CrossRef]

55. Thomas, R.J.; Lascano, C.E. The benefits of forage legumes for livestock production and nutrient cycling in pasture and agropastoral systems of acid soils savannas of Latin America. In *Livestock and Sustainable Nutrient Cycling in Mixed Farming Systems of Sub-Sahara Africa*; Powell, J.M., Fernandez-Rivera, S., Williams, T.O., Renard, C., Eds.; ILCA: Addis Ababa, Ethiopia, 1995; pp. 277–291.

56. Rao, I.M.; Kerridge, P.C.; Macedo, M.C.M. Nutritional requirements of *Brachiaria* and adaptation to acid soils. In *Brachiaria: Biology, Agronomy, and Improvement*; Miles, J.W., Maass, B.L., do Valle, C.B., Eds.; CIAT: Cali, Colombia, 1996; pp. 53–71.

57. Rao, I.; Kerridge, P. Mineral nutrition of forage Arachis. In *Bioogy and Agronomy of Forage Arachis*; Kerridge, P., Hardy, B., CIAT: Cali, Colombia, 1994; pp. 71–83.

58. Cantarutti, R.B.; Tarre, R.; Macedo, R.; Cadisch, G.; Rezende, C.D.; Pereira, J.M.; Braga, J.M.; Gomide, J.A.; Ferreira, E.; Alves, B.J.R.; et al. The effect of grazing intensity and the presence of a forage legume on nitrogen dynamics in *Brachiaria* pastures in the Atlantic forest region of the south of Bahia, Brazil. *Nutr. Cycl. Agroecosyst.* **2002**, *64*, 257–271. [CrossRef]

59. Rao, I.M.; Borrero, V.; Ricaurte, J.; Garcia, R. Adaptive attributes of tropical forage species to acid soils. V. Differences in phosphorus acquisition from less available inorganic and organic sources of phosphate. *J. Plant Nutr.* **1999**, *22*, 1175–1196. [CrossRef]

60. Oberson, A.; Bünemann, E.K.; Friesen, D.K.; Rao, I.M.; Smithson, P.C.; Turner, B.L.; Frossard, E. Improving phosphorus fertility in tropical soils through biological interventions. In *Biological Approaches to Sustainable Soil Systems*; Uphoff, N., Ball, A.S., Fernandes, E., et al., Eds.; CRC Press: Boca Raton, FL, USA, 2006; pp. 531–546.

61. Oberson, A.; Friesen, D.K.; Tiessen, H.; Morel, C.; Stahel, W. Phosphorus status and cycling in native savanna and improved pastures on an acid low-P Colombian Oxisol. *Nutr. Cycl. Agroecosyst.* **1999**, *55*, 77–88. [CrossRef]

62. Bünemann, E.; Smithson, P.C.; Jama, B.; Frossard, E.; Oberson, A. Maize productivity and nutrient dynamics in maize-fallow rotations in western Kenya. *Plant Soil* **2004**, *264*, 195–208. [CrossRef]

63. Rao, I.M.; Barrios, E.; Amezquita, E.; Friesen, D.K.; Thomas, R.; Oberson, A.; Singh, B.R. Soil phosphorus dynamics, acquisition and cycling in crop-pasture-fallow systems in low fertility tropical soils of Latin America. In *Modelling Nutrient Management in Tropical Cropping Systems*; Delve, R.J., Probert, M.E., Eds.; Australian Center for International Agricultural Research (ACIAR): Canberra, Australia, 2004; Volume 114, pp. 126–134.

64. Nakamura, T.; Miranda, C.H.B.; Ohwaki, Y.; Valéio, J.R.; Kim, Y.; Macedo, M.C.M. Characterization of nitrogen utilization by *Brachiaria* grasses in Brazilian savannas (Cerrados). *Soil Sci. Plant Nutr.* **2005**, *51*, 973–979. [CrossRef]

65. O'Leary, M.H. Carbon Isotopes in Photosynthesis: Fractionation techniques may reveal new aspects of carbon dynamics in plants. *BioScience* **1988**, *38*, 328–336. [CrossRef]

66. De Moraes, J.F.L.; Volkoff, B.; Cerri, C.C.; Bernoux, M. Soil properties under Amazon forest and changes due to pasture installation in Rondônia, Brazil. *Geoderma* **1996**, *70*, 63–81. [CrossRef]

67. Craine, J.M.; Brookshire, E.N.J.; Cramer, M.D.; Hasselquist, N.J.; Koba, K.; Marin-Spiotta, E.; Wang, L. Ecological interpretations of nitrogen isotope ratios of terrestrial plants and soils. *Plant Soil* **2015**, *396*, 1–26. [CrossRef]

Publisher's Note: MDPI stays neutral with regard to jurisdictional claims in published maps and institutional affiliations.

Article

Functional Diversity Effects of Vegetation on Runoff to Design Herbaceous Hedges for Sediment Retention

Léa Kervroëdan [1,2,*], Romain Armand [1,2], Mathieu Saunier [2] and Michel-Pierre Faucon [1]

[1] AGHYLE, UP 2018.C101, SFR Condorcet FR CNRS 3417, UniLaSalle, 60026 Beauvais, France
[2] AREAS, 76460 St Valéry en Caux, France
* Correspondence: lea.kervroedan@unilasalle.fr

Received: 12 February 2020; Accepted: 27 March 2020; Published: 31 March 2020

Abstract: Background: Functional diversity effects on ecosystem processes, like on soil erosion, are not fully understood. Runoff and soil erosion in agricultural landscapes are reduced by the hydraulic roughness (HR) of vegetation patches, which furthers sediment retention. Vegetation with important stem density, diameters, leaf areas, and density impact the HR. A functional structure composed of these negatively correlated traits involved in the increase of the HR would constitute a positive effect of the functional diversity. Methods: Runoff simulations were undertaken on four mono-specific and two multi-specific communities, using herbaceous plant species from North-West Europe, presenting six contrasting aboveground functional traits involved in the HR increase. Results: An effect of dominant traits in the community was found on the HR, identified as the community-weighted leaf density. The non-additive effect of functional diversity on the HR could be explained by the presence of species presenting large stems in the communities with high functional diversity. Conclusion: We argued that functional diversity effect on the HR could change due to idiosyncratic effects of the plant traits, which would be influenced by soil properties, phylogeny diversity, and plant species interactions. These findings constitute an advancement in the understanding of plant trait assemblage on runoff and soil erosion processes.

Keywords: functional diversity; hydraulic roughness; herbaceous vegetation; leaf and stem functional traits; plant–runoff interaction; soil erosion control

1. Introduction

Natural ecosystem processes are driven by the plant functional traits in vegetation communities [1,2]. Plant functional diversity, defined as "the value, range and relative abundance of plant functional traits in a given ecosystem" [3,4], can play a major role in ecosystem functioning and in supplying ecosystem services [2,5–7]. Studies have focused on the effects of functional diversity on ecosystem processes to understand if these effects were due to a dominant species composing the community or due to its functional diversity [3,6,8,9]. The mass ratio hypothesis stipulates that ecosystem processes would be driven by the traits of the most abundant species in the community, characterised as dominant species, and is represented by the community-weighted mean traits [10,11]. On the contrary, the diversity hypothesis specifies that ecosystem processes are driven by the trait diversity composing the plant community, inducing complementarity effects among species [7,12]. Higher dissimilarity of traits in a community would lead to a more complete use of resources and, thus, to a more important plant productivity [9,13], as well as a stronger impact on less studied plant–soil related processes like soil erosion, through an increase of hydraulic roughness and sediment retention [14].

The effect of plant functional diversity on soil erosion processes have recently appealed to the interest of the scientific community that studies plant–soil erosion processes, although the results of the functional diversity effects are contentious [14–16]. However, these studies focused on the effects of functional diversity of root traits of non-herbaceous communities, on soil stabilization and resistance in mountainous or semi-arid vegetation, which are community structures selected under the erosion processes, specific to these soil and climatic contexts [14,17,18]. In landscapes where annual crop fields on loamy soils represent an important area, runoff and linear soil erosion are mainly reduced by the hydraulic roughness in small vegetation patches [19]. Hydraulic roughness is defined as frictional resistance due to the contact of runoff with the vegetation [20–24]. Hydraulic roughness of herbaceous vegetation furthers sediment retention by reducing flow velocity; however, it is highly variable depending on the plant species and traits [21,24–29]. A positive relationship between the aboveground biomass and hydraulic roughness was highlighted, as an increase in the biomass productivity would further hydraulic roughness and sediment retention [25]. The aboveground functional traits that directly impact the hydraulic roughness for erosional events found under temperate climates were identified by Kervroëdan et al. [26]—herbaceous vegetation with important leaf density, leaf area, stem diameter, and stem projected area (stem area toward the flow direction) were found to be the most efficient in increasing hydraulic roughness. Nonetheless, these results emphasized the effects of negatively correlated trait combinations (i.e., leaf density and area) involved in hydraulic roughness increase, which suggested that communities with a high functional diversity would reach the best trade-off to maximize the vegetation effects on hydraulic roughness. A primary analysis was undertaken on the effects of aboveground functional trait divergence in herbaceous vegetation on the hydraulic roughness and sediment retention for processes occurring in loamy agricultural soils, emphasizing a dominant effect of the community-weighted traits in vegetation [16]. However, these results highlighted the need to deepen the understanding of the effects of functional diversity on runoff and linear soil erosion processes, by integrating a functional diversity gradient within the tested conditions.

In this study on trait-based plant ecohydrology, we aimed to examine the effects of functional diversity on hydraulic roughness, based on a functional diversity gradient, by using four monospecific and two multi-specific communities. We predicted that functional diversity increased the hydraulic roughness, by exerting a synergistic effect. The stem and leaf traits involved in the hydraulic roughness increase (stem diameter and projected area; leaf density, area, and specific area) would present a complementarity in the space-use, representing a competitive balance among the species in the community. The communities with a high differentiation degree among these traits will, thus, use the aboveground space more efficiently and lead to an increase in hydraulic roughness.

2. Materials and Methods

2.1. Plant Material

Four indigenous plant species from North-West Europe that presented selected functional types, contrasting aboveground functional traits, and that were involved in the increase of hydraulic roughness (leaf—area and density; stem—projected area, diameter, and density) [26], were used in the present study. The chosen plant species presented a minimal vegetative height within the range of 20 and 60 cm, in order to limit competition for light and ensure a uniform development of each species in the plots. Three replicates of monospecific and two multi-specific communities of *Carex flacca* Schreb. I, *Tanacetum vulgare* L. (T), *Festuca arundinacea* Schreb. (F), and *Phalaris arundinacea* L. (P), with contrasting traits were tested (Figure 1A). There were two types of multi-specific communities, one composed of *C. flacca*, *T. vulgare*, and *F. arundinacea* (CTF); and the other with *C. flacca*, *T. vulgare*, *F. arundinacea*, and *P. arundinacea* (CTFP). Each species was collected in their natural habitat in March 2016 (i.e., three months before the experiments), to ensure the creation of densely planted plots with grown individuals. The experiments were ex-situ and the plants were planted in a 60 × 30 × 15 cm wooden frame with a grid fence at the bottom to allow the roots' development. The multi-specific plots were covered with the same proportion of each species, which were placed such that patches of the same species would not be in contact.

Figure 1. Experimental design: (**A**) Three levels of plant functional diversity with four monospecific and two multi-specific conditions; and (**B**) runoff simulator set-up used for the experiments and water level measurement. The photos of the plots were taken 1.5 month prior to the experiments. C – *Carex flacca*, T—*Tanacetum vulgare*, F—*Festuca arundinacea* and P—*Phalaris arundinacea*; CTF—*Carex flacca + Tanacetum vulgare + Festuca arundinacea*; CTFP—*Carex flacca + Tanacetum vulgare + Festuca arundinacea + Phalaris arundinacea*.

2.2. Plant Functional Traits Measurements

Six aboveground functional plant traits (stem—density, diameter, and projected area; leaf—density and area) that are known to increase the hydraulic roughness [29], were measured on three levels along the stem—between 0 and 5 cm, 0 and 10 cm, and 0 and 20 cm. Guidelines from Pérez-Harguindeguy et al. [30] were followed regarding the sampling collection, samples conservation, and analyses methods. As analyses could not be performed directly after sampling, the leaves were stored in sealed bags with moist tissue until measurements, and were then dried at 70 °C for 72 h.

As the plots presented a homogeneous plant cover, all traits measurements were carried out within one quadrat (10 by 10 cm) per monospecific community plot and one quadrat per species within the multi-specific community plots. The determination of stem density (stem dm^{-2}) included plant stems, as well as pseudoculms for the sedges species (Cyperaceae) and tillers for the grass species, which here are considered to have the same functional effect on hydraulic roughness, as the stems. The stem diameter (mm) was measured on three representative stems and was used to determine the stem projected area (mm^2) of each stem, using the rectangle area formula. The leaf area (mm^2) and the specific leaf area (mm^2 mg^{-1}) were estimated from six representative leaves, which were scanned using a 600 dpi resolution and the images were processed using Gimp 2.8. The aboveground biomass (g) was entirely removed, dried at 70 °C for 72 h and weighted.

Furthermore, the density of traits within the quadrat was calculated through the product of (1) the leaf traits by the leaf density and (2) the stem traits by the stem density. These densities and the traits were dissociated with a "D" in front of the trait names.

2.3. Characterisation of the Community Functional Structure

Both multi-specific communities were created with an equal abundance of each species. The CTF communities accounted for 33% of each species (*C. flacca, T. vulgare,* and *F. arundinacea*) and the CTFP communities accounted for 25% of each species (*C. flacca, T. vulgare, F. arundinacea,* and *P. arundinacea*).

The community-weighted traits (CWT) were calculated, for each trait in both multi-specific communities, as the mean trait value after each trait was weighted by the abundance of each species composing the community [11,31]:

$$CWT_i = \sum_{k=1}^{n_i} A_{k,i} T_{k,i}, \tag{1}$$

where CWT_i is the community-weighted value of the trait in the community i, $A_{k,i}$, and $T_{k,i}$ are, respectively, the relative abundance and the trait value of the species k in the community i and n_i is the number of species in community i.

The functional variance (FD_{var}) was calculated for each trait in both multi-specific communities. FD_{var} represents the variance of the trait values of the species in the community [32]:

$$FD_{var} = \frac{2}{\pi} \arctan[5 \sum_{i=1}^{N} [(\ln C_i - \overline{\ln x})^2 A_i, \tag{2}$$

where C_i is the value of the trait i, $\overline{\ln x}$ is the weighted logarithmic mean of the trait and A_i is the abundance of the species with trait i [32,33].

The Rao's quadratic entropy (FD_Q) [34], a multidimensional index of the functional diversity to reflect the diversity hypothesis, was also used to characterize the functional divergence within the multi-specific conditions. It is a generalized form of the diversity Simpson index [35] and combines a measure of the pairwise functional differences between species and the relative abundance of the species [36]. The FD_Q was determined using the package FD in R (version 3.3.2).

2.4. Hydraulic Measurements

The effect of the functional diversity on the hydraulic roughness was measured using a runoff simulator that recreated a flow at set discharges in controlled conditions [37]. The flow discharge was monitored through Venturi channels (flow range of 0.06–6 L s^{-1}) and ultrasound probes measuring the water level in the channels (±1.26 mm), located in the upper and lower parts of the simulator. This system was manufactured by ISMA, France [37]. Measures of the hydrological processes were carried out in the central part of the simulator, a channel area which consisted of two 5.40 m galvanized iron sheets buried 60 cm away from each other on a 5% levelled slope. The entire channel area was waterproofed using a tarpaulin to avoid water losses during the experiment, such as leaks and infiltration in the ground. The plot was located 4 m away from the head of the channel in a 17 cm deep rectangular hole to level the plants with the channel (Figure 1B). To measure the channel topography and the backwater level in front of the plants, five spacers were placed upstream of the plot, from approximately 1.46 m from the channel head, every 0.75 m. Each spacer was levelled and its elevation was measured to use them as elevation-known baselines for the water level measurements.

In order to investigate the behaviour of the plant communities towards processes occurring more or less frequently, four discharges were used: 2, 4, 8, and 11 L s^{-1} m^{-1} at ±7% (observed approximately every 0.5, 1, 2, and 5 years, respectively, in 5 ha catchments with a 5 m-wide thalweg [37]). Discharges were continuously monitored through both upstream and downstream flowmeters. When the upstream and downstream discharges were equivalent, the water levels were measured as the perpendicular distance between the bottom of the spacer and the top of the water flow (Figure 1B), using the closest spacer upstream of the plot. Seven vertical water profiles were made per discharge per plot, one every 10 cm along the spacer from the edges of the channel.

The unit stream power (USP) was used here as a proxy to characterize the hydraulic roughness, being often used as a sediment transport capacity index [38,39], as it represents the "energy dissipation per unit of time and per unit of weight of the flow" [39]:

$$USP = VS,\qquad(3)$$

where USP (m s^{-1}) depends on V, the mean velocity of the flow (m s^{-1}), and S, the slope of the channel (m m^{-1}). The USP is negatively related to the hydraulic roughness—the lower it is, the higher is the hydraulic roughness. The Manning n coefficients and the water levels data at each discharge for each condition are provided as supplementary data (Table S1).

2.5. Data Analysis

Mann Whitney and T tests were conducted on the functional variance for each trait to analyze the variation of the functional structure between the two multi-specific communities. After regrouping the data under the categories "monospecific" and "multi-specific", Mann Whitney analyses were performed on the USP data to compare (1) the mean value of monospecific communities with the multi-specific communities and (2) both multi-specific communities. The monospecific P community data were excluded from the analysis for the comparison with the three-species communities (CTF). ANOVA and Tukey post-hoc tests were then computed on the USP data to examine the differences between each community category to understand if one species/community had more impact on the USP than another.

ANOVA and Kruskal-Wallis analyses, as well as respective post-hoc tests Tukey and Mann-Whitney were used on the traits, the community-weighted traits and the biomass data, according to the normality of the data, to examine the differences in trait and community-weighted trait compositions between the communities. Moreover, after combining the data under "monospecific" and "multi-specific" categories, Mann Whitney analyses were carried out on the biomass and the community-weighted trait data, to compare the mean value of the monospecific communities with the multi-specific communities. Regarding the analysis of the CTF communities, the monospecific P communities were removed from the data.

All statistical analyses were computed using the statistical software R (version 3.3.2).

3. Results

3.1. Variation of Community-Weighted Trait and the Unit Stream Power

According to the results from the comparison analyses between each community on the USP, only the *F. arundinacea* in monoculture presented a significant difference to the other communities (Figure 2). Similar results were found at each tested discharge (Table 1). There were no significant differences between the other communities studied. The results on the aboveground biomass did not show any differences between the mean of the monospecific with the multi-specific communities (Figure 3). The analyses comparing the biomass productivity of all the conditions showed a significant difference between *C. flacca* and *T. vulgare* (with a trend between *C. flacca* and all other communities, as well as between *T. vulgare* and *F. arundinacea*, as the *p*-values from the Mann-Whitney tests were 0.057). Regarding the community-weighted trait analysis, all ANOVA/Kruskal-Wallis tests were found to be significant (Table 2). Except for the CW leaf area (CW-LA), CW leaf density (CW-LD; 0–10 cm), CW leaf density (CW-LD; 0–20 cm), and CW density-weighted SLA (CW-DSLA; 0–10 cm), both multi-specific communities showed no significant difference from any monospecific community. CW density-weighted stem projected (CW-DSA) and CW density-weighted stem diameter (CW-DSDm), for all levels along the stem, showed only a difference of *F. arundinacea*, within the monospecific communities. Differences between *F. arundinacea* and the multi-specific communities were found for the CW leaf density (0–20 cm) (with both multi-specific communities) and CW leaf density (0–10 cm) (only with CTFP).

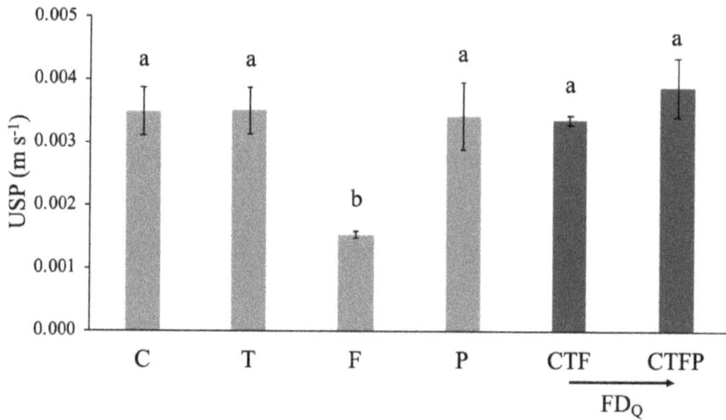

Figure 2. Differences between each community, at discharge 2 L s^{-1} m^{-1}. The bars represent the mean ± standard error. The letters represent the significant differences between each community according to the Tukey post-hoc tests. The arrow represents the increase of the functional diversity between both multi-specific conditions (with Rao's quadratic entropy (FDQ)).

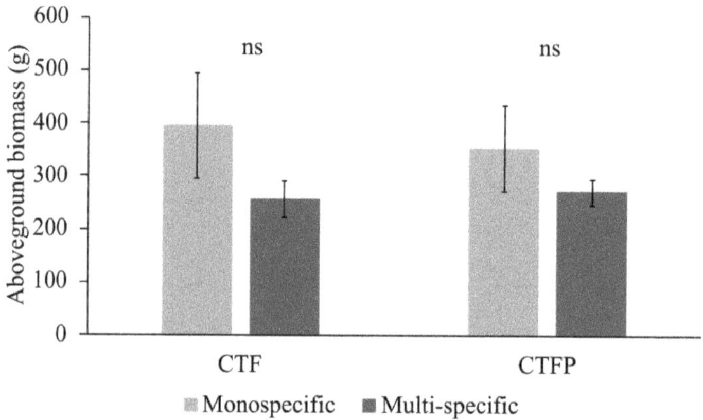

Figure 3. Differences in biomass productivity between the mean of the monospecific communities and the multi-specific communities. The bars represent the mean ± standard error. Significance level: ns = not significant.

Table 1. Differences in unit stream power for each community tested at each discharge.

Discharges	C	T	F	P	CTF	CTFP	ANOVA
Q1 = 2 L s^{-1} m^{-1}	0.0035 (±0.0008) [a]	0.0035 (±0.0007) [a]	0.0015 (±0.0001) [b]	0.0034 (±0.0011) [a]	0.0033 (±0.0001) [a]	0.0039 (±0.0008) [a]	5.29 **
Q2 = 4 L s^{-1} m^{-1}	0.0046 (±0.0007) [a]	0.0051 (±0.0009) [a]	0.0021 (±0.0001) [b]	0.0046 (±0.0011) [a]	0.0044 (±0.0002) [a]	0.0053 (±0.0008) [a]	8.8 ***
Q3 = 8 L s^{-1} m^{-1}	0.0066 (±0.001) [a]	0.0083 (±0.0013) [a]	0.0032 (±0.0002) [b]	0.0067 (±0.0013) [a]	0.0064 (±0.0004) [a]	0.0077 (±0.0015) [a]	10.92 ***
Q4 = 11 L s^{-1} m^{-1}	0.0078 (±0.0011) [a]	0.0105 (±0.0015) [a]	0.0038 (±0.0001) [b]	0.0079 (±0.0016) [a]	0.0077 (±0.0006) [a]	0.009 (±0.0017) [a]	12.64 ***

Data are unit stream power (USP) mean values (±standard deviation) and the results of the statistical tests F in ANOVA. C—*Carex flacca*; T—*Tanacetum vulgare*; F—*Festuca arundinacea*; P—*Phalaris arundinacea*; CTF—*Carex flacca + Tanacetum vulgare + Festuca arundinacea*; and CTFP—*Carex flacca + Tanacetum vulgare + Festuca arundinacea + Phalaris arundinacea*. The letters represent the significant differences between the communities, according to the Tukey post-hoc tests. Significance levels: *** = $p < 0.001$; ** = $p < 0.01$; * = $p < 0.05$; ns = not significant.

Table 2. Differences in community-weighted traits (CW) for each community tested.

Community-Weighted Traits	C	T	F	P	CTFP	CTF	ANOVA /Kruskal-Wallis
CW-LA (mm^2)	1575.8 (±144.6) [a]	4168.5 (±1049.5) [b]	3130 (±298.4) [bc]	3054.4 (±331.1) [bc]	3116.2 (±605.6) [bc]	2715.9 (±224.8) [ac]	9.28 ***
CW-LD5 (dm^{-2})	96.3 (±15.6) [a]	12 (±2.6) [b]	193.1 (±73.3) [c]	46.6 (±24.5) [a]	91.2 (±7.1) [abc]	111.2 (±21.8) [abc]	**18.99 ****
CW-LD10 (dm^{-2})	148.8 (±30.8) [a]	20.8 (±1.5) [b]	280.3 (±79.6) [c]	74.9 (±43.1) [ab]	154.8 (±37.8) [a]	171.7 (±31.4) [ac]	15.53 ***
CW-LD20 (dm^{-2})	236.5 (±43.4) [a]	38.8 (±2.6) [b]	423.9 (±61.4) [c]	108.5 (±61.4) [bd]	214.8 (±79.7) [ad]	212.8 (±40.2) [ad]	24.89 ***
CW-DLA5 (mm^2 dm^{-2})	153,175 (±35,395) [a]	50,527 (±20,693) [b]	599,417 (±210,360) [c]	147,223 (±90,166) [ab]	229,484 (±25,910) [abc]	251,565 (±59,698) [abc]	**17.96 ****
CW-DLA10 (mm^2 dm^{-2})	236,709 (±61,527) [a]	86,420 (±23,545) [b]	868,938 (±216,666) [c]	237,161 (±156,858) [a]	406,719 (±121,334) [abc]	389,669 (±77,960) [abc]	**17.63 ****
CW-DLA20 (mm^2 dm^{-2})	376,697 (±93,429) [a]	163,234 (±48,516) [b]	1,325,651 (±212,015) [c]	343,506 (±223,961) [ab]	573,873 (±251,557) [abc]	474,577 (±79,275) [abc]	**15.91 ****
CW-DSA5 (mm^2 dm^{-2})	4675.3 (±1258.3) [a]	2263.4 (±593.7) [a]	19,926 (±7469.9) [b]	3880.1 (±2061) [a]	6088.6 (±1345) [ab]	5579.4 (±1499.8) [ab]	**15.48 ****
CW-DSA10 (mm^2 dm^{-2})	9226.5 (±2471.1) [a]	4412.9 (±1116.5) [a]	40,673 (±15,603) [b]	7458.3 (±3858.2) [a]	12,040 (±2561.4) [ab]	11,169 (±2975.2) [ab]	**15.95 ****
CW-DSA20 (mm^2 dm^{-2})	16,841 (±4587.3) [a]	8739.8 (±2140.3) [a]	82,565 (±30,896) [b]	14,356 (±7515.9) [a]	23,190 (±5038.9) [ab]	21,497 (±5331.6) [ab]	**16.09 ****
CW-DSDm5 (mm dm^{-2})	93.5 (±25.2) [a]	45.3 (±11.9) [a]	398.5 (±149.4) [b]	77.6 (±41.2) [a]	121.8 (±26.9) [ab]	111.6 (±30) [ab]	**15.48 ****
CW-DSDm10 (mm dm^{-2})	92.3 (±24.7) [a]	44.1 (±11.2) [a]	406.7 (±156) [b]	74.6 (±38.6) [a]	120.4 (±25.6) [ab]	111.7 (±29.8) [ab]	**15.95 ****
CW-DSDm20 (mm dm^{-2})	91.5 (±24) [a]	43.7 (±10.7) [a]	412.8 (±154.5) [b]	71.8 (±37.6) [a]	119.2 (±25.4) [ab]	110.6 (±27.1) [ab]	**15.95 ****

Data are CW mean values (±standard deviation) and results of F (ANOVA) and Chi-squared (Kruskal-Wallis, in bold) statistical tests. C—*Carex flacca*; T—*Tanacetum vulgare*; F—*Festuca arundinacea*; P—*Phalaris arundinacea*; CTF—*Carex flacca + Tanacetum vulgare + Festuca arundinacea*; CTFP—*Carex flacca + Tanacetum vulgare + Festuca arundinacea + Phalaris arundinacea*; LA—leaf area; LD—leaf density; DLA—leaf area density; DSA—stem area density; and DSDm—stem diameter density. The letters represent the significant differences between the communities, according to the Tukey and Mann-Whitney post-hoc tests. Significance levels: *** = $p < 0.001$; ** = $p < 0.01$; * = $p < 0.05$; ns = not significant.

3.2. Variation of Functional Diversity within the Communities

The results on the functional variance showed significant differences and a higher variance of the traits in CTFP than in CTF, except for the leaf area (LA) and the density-weighted leaf area (DLA; 0–5 cm) (Table 3). The values of FD_Q found for the community CTF was 10 and 11.25 for the community CTFP.

Table 3. Summary of the functional diversity variance FDvar for both multi-specific communities and results from the T and Mann-Whitney tests.

Traits	CTF	CTFP	t/W
LA (mm^2)	0.7454 (±0.0836)	0.7614 (±0.0735)	−0.24864 ns
LD5 (dm^{-2})	0.9855 (±0.0009)	0.9883 (±0.0006)	−4.5579 *
LD10 (dm^{-2})	0.9879 (±0.0008)	0.9905 (±0.0007)	−4.1791 *
LD20 (dm^{-2})	0.9889 (±0.0009)	0.9917 (±0.001)	−3.6474 *
DSA5 (mm^2 dm^{-2})	0.9967 (±0.0002)	0.9974 (±0.0002)	−4.9464 *
DSA10 (mm^2 dm^{-2})	0.9972 (±0.0002)	0.9979 (±0.0001)	−5.5758 **
DSA20 (mm^2 dm^{-2})	0.9976 (±0.0001)	0.9982 (±0.0001)	−6.6792 **
DSDm5 (mm dm^{-2})	0.9878 (±0.0013)	0.9903 (±0.0009)	−2.7224 °
DSDm10 (mm dm^{-2})	0.9878 (±0.0012)	0.9902 (±0.0009)	−2.8 °
DSDm20 (mm dm^{-2})	0.9878 (±0.001)	0.9902 (±0.0009)	−3.0036 *
DLA5 (mm^2 dm^{-2})	0.9986 (±0.00003)	0.999 (±0.00003)	**0 ns**
DLA10 (mm^2 dm^{-2})	0.9987 (±0.00002)	0.9991 (±0.0001)	−11.108 **
DLA20 (mm^2 dm^{-2})	0.9988 (±0.00003)	0.9992 (±0.0001)	−8.8503 **

Data are FDvar mean values (±standard deviation) and t values of T tests and the W value of Mann-Whitney (in bold) test. CTF—*Carex flacca* + *Tanacetum vulgare* + *Festuca arundinacea*; CTFP—*Carex flacca* + *Tanacetum vulgare* + *Festuca arundinacea* + *Phalaris arundinacea*; LA—leaf area; LD—leaf density; DLA—leaf area density; DSA—stem area density; and DSDm—stem diameter density. Significance levels: *** = $p < 0.001$; ** = $p < 0.01$; * = $p < 0.05$; ° = $p < 0.1$; ns = not significant.

3.3. Variation of Functional Diversity and the Unit Stream Power

The comparison analysis of the USP values for the monospecific communities and for the multi-specific communities, using the Kruskal-Wallis tests, showed no significant differences for any of the combinations tested (Figure 4). Moreover, no difference was observed between both multi-specific communities and the USP did not show a decrease of its value with an increase of functional diversity in the communities (Figure 4). Similar results were found through all tested discharges.

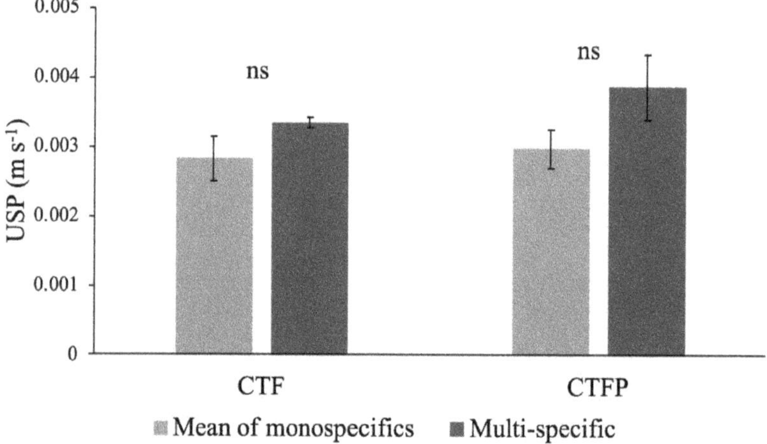

Figure 4. Differences between the mean of the monospecific and the multi-specific conditions at the discharge 2 L s^{-1} m^{-1}. The bars represent the mean ± standard error. Significance level: ns = not significant.

4. Discussion

4.1. Non-Additive Effect of the Functional Diversity on Hydraulic Roughness

The hydraulic roughness is mainly increased by aboveground plant functional traits in herbaceous plant communities, inducing sediment retention. Associating contrasting traits within a plant community would lead to a better trade-off in the traits values and would more efficiently increase the hydraulic roughness and sediment retention [26]. An increase of functional variance was observed for most of the traits with an increase of functional diversity in the multi-specific conditions. These differences in functional diversity observed among the communities should have led to an additive effect on hydraulic roughness. Indeed, the complementarity hypothesis stipulates that the diversity in trait attributes in a community influences the ecosystem processes, by inducing synergetic effects of functional diversity (i.e., complementarity or facilitation effects) among the coexisting species [7,40]. However, the results did not show any synergetic effect of the functional diversity on hydraulic roughness, in the presence of contrasting traits. This could be justified by the absence of complementarity between the traits, as there were no significant difference between the mean of the monospecific and the multi-specific communities, nor between both multi-specific communities presenting different functional diversities (FD$_Q$). These findings highlighted the dominant effect of the community-weighted traits of specific species in the communities, on runoff. The vegetation effect on the hydraulic roughness was, thus, driven by the traits of one or more species in the community, which presented a dominant effect on the process compared to the traits of other species [6,10,11]. The results emphasized the effects of the community-weighted trait of *F. arundinacea* on hydraulic roughness being the only species that was significantly different from the other communities, which would be led by the different CW leaf densities (0–20 cm and 0–10 cm), which is a major trait that is known to positively impact hydraulic roughness [26]. While similar effects of the dominance of community-weighted traits on soil retention have been observed in mixtures of different functional types [15] and herbaceous vegetation [16], additive effects of functional diversity on erosion rates were also shown [14]. These opposing effects of functional diversity on the hydraulic roughness and erosion rates could be explained by the idiosyncratic effects, which represent the contrasting effects of functional diversity affected by species and trait interactions, and soil–plant interactions. Idiosyncratic effects could be influenced by the functional structure within the communities [15]. Within the chosen species, *T. vulgare* presented larger stem diameters than other species, but if the proportion of large stem diameters was too high in the community, preferential flow paths would be taken by water, limiting the effects of denser individuals, which would impact the effect of functional diversity on hydraulic roughness [15,19]. This hypothesis is consistent with results showing that the presence of functional types with larger stem diameters would influence the water path within a vegetation and, thus, displayed no effect of community functional diversity [15]. These findings are also coherent with primary results obtained on herbaceous vegetation, showing the negative influence of plants with low stem density, large diameters, and low leaf density, by reducing the hydraulic roughness and sediment retention within multi-specific communities [16]. The perspective is to study a gradient of functional structure, using a variation of abundances of large stem diameters, to confirm and show the idiosyncratic effects of functional diversity on water flow.

4.2. Implication to Design Herbaceous Hedges for Sediment Retention

These findings constitute an advancement in the understanding of community functional structure effects on runoff and hydraulic roughness. The results highlighted that the presence of high leaf density species would tend to dominate the community effects on hydraulic roughness and, thus, sediment retention. This response was found to be independent from the intensity of the event occurrence, as similar results were found for all tested discharges.

In order to model the effects of multi-specific plant communities on runoff and sediment retention processes, the community-weighted mean value of traits could be implemented, as, depending on the functional structure of the community, a non-additive effect of functional diversity could be found.

The integration of the community-weighted traits into modeling vegetation effects would allow to take this dominance effect into account and modeled the community overall effects on runoff and soil erosion in herbaceous hedges that were essentially involved in sediment retention.

Moreover, an absence of negative effect of the functional diversity was found in this study; however, a high competition within the herbaceous hedge due to the species combination could lead to a decrease of stem and leaf densities in the plant community and, thus, to a negative effect of functional diversity on hydraulic roughness. Using combinations of species involved in hydraulic roughness, within the hedge design for increased soil erosion control, is recommended, as long as the competitiveness of the plants (e.g., same range of vegetative height) is taken into account in the species selection process. Indeed, using multi-specific communities would create multi-functional ecosystems that could offer other ecosystem services, such as biodiversity conservation through the creation of new habitats and the enhancement of ecological connectivity, to mitigate biodiversity erosion [41].

5. Conclusions

This ecohydrology study allowed to identify the relationship between the aboveground functional structure and the hydraulic roughness. The results showed that there was no influence of the functional diversity on the hydraulic roughness but rather an effect of the dominant traits was observed in the community, identified as the community-weighted leaf densities. The absence of functional diversity effect on hydraulic roughness would be explained through the idiosyncratic effects of the traits. A new perspective would be to study the functional diversity effects on a wider diversity gradient of candidate species that are involved in the increase of hydraulic roughness, in order to explain the contrasting results found for the functional diversity effects on sediment retention and runoff processes.

Supplementary Materials: The following are available online at http://www.mdpi.com/1424-2818/12/4/131/s1. Table S1: Differences in Manning n coefficient and water levels for each community tested at each discharge.

Author Contributions: Conceptualization, L.K., M.-P.F. and R.A.; Methodology, L.K., M.-P.F. and R.A.; Validation, L.K., M.-P.F. and R.A.; Formal Analysis, L.K.; Resources, L.K., M.-P.F., M.S. and R.A.; Data Curation, L.K. and M.S.; Writing—Original Draft Preparation, L.K.; Writing—Review & Editing, L.K., M.-P.F. and R.A.; Supervision, M.-P.F. and R.A. All authors have read and agreed to the published version of the manuscript.

Funding: This research received no external funding.

Acknowledgments: The authors thank the funders of this study: Agence de l'Eau Seine-Normandie (Seine-Normandie Catchment Agency), Région Normandie (Normandie council), and ANRT (National Association for Technological Research). Special acknowledgements are given to Yves le Bissonnais and Freddy Rey for their scientific comments on the methods of our study and to Jean-François Ouvry and M. Jean-Baptiste Richet for their insightful comments and technical advices.

Conflicts of Interest: The authors declare no conflict of interest. The funders had no role in the design of the study; in the collection, analyses, or interpretation of data; in the writing of the manuscript, or in the decision to publish the results.

References

1. Lavorel, S.; McIntyre, S.; Landsberg, J.; Forbes, T.D.A. Plant functional classifications: From general groups to specific groups based on response to disturbance. *Trends Ecol. Evol.* **1997**, *12*, 474–478. [CrossRef]
2. Cadotte, M.W.; Carscadden, K.; Mirotchnick, N. Beyond species: Functional diversity and the maintenance of ecological processes and services. *J. Appl. Ecol.* **2011**, *48*, 1079–1087. [CrossRef]
3. Díaz, S.; Lavorel, S.; Chapin, F.S.; Tecco, P.A.; Gurvich, D.E.; Grigulis, K. Functional Diversity—At the Crossroads between Ecosystem Functioning and Environmental Filters. In *Terrestrial Ecosystems in a Changing World*; Canadell, J.G., Pataki, D.E., Pitelka, L.F., Eds.; Springer Berlin Heidelberg: Berlin/Heidelberg, Germany, 2007; pp. 81–91, ISBN 978-3-540-32729-5.
4. Tilman, D. Functional Diversity. In *Encyclopedia of Biodiversity*; Levin, S.A., Ed.; Elsevier: Amsterdam, The Netherlands, 2001; pp. 109–120, ISBN 978-0-12-226865-6.
5. Faucon, M.-P.; Houben, D.; Lambers, H. Plant functional traits: Soil and ecosystem services. *Trends Plant Sci.* **2017**, *22*, 385–394. [CrossRef] [PubMed]

6. Lavorel, S.; Garnier, E. Predicting changes in community composition and ecosystem functioning from plant traits: Revisiting the Holy Grail. *Funct. Ecol.* **2002**, *16*, 545–556. [CrossRef]
7. Tilman, D.; Knops, J.; Wedin, D.; Reich, P.; Ritchie, M.; Siemann, E. The influence of functional diversity and composition on ecosystem processes. *Science* **1997**, *277*, 1300–1302. [CrossRef]
8. Song, Y.; Wang, P.; Li, G.; Zhou, D. Relationships between functional diversity and ecosystem functioning: A review. *Acta Ecol. Sin.* **2014**, *34*, 85–91. [CrossRef]
9. Cadotte, M.W. Functional traits explain ecosystem function through opposing mechanisms. *Ecol. Lett.* **2017**, *20*, 989–996. [CrossRef]
10. Grime, J.P. Benefits of plant diversity to ecosystems: Immediate, filter and founder effects. *J. Ecol.* **1998**, *86*, 902–910. [CrossRef]
11. Díaz, S.; Lavorel, S.; de Bello, F.; Quetier, F.; Grigulis, K.; Robson, T.M. Incorporating plant functional diversity effects in ecosystem service assessments. *Proc. Natl. Acad. Sci. USA* **2007**, *104*, 20684–20689. [CrossRef]
12. Petchey, O.L. Integrating methods that investigate how complementarity influences ecosystem functioning. *Oikos* **2003**, *101*, 323–330. [CrossRef]
13. Loreau, M.; Hector, A. Partitioning selection and complementarity in biodiversity experiments. *Nature* **2001**, *412*, 72–76. [CrossRef] [PubMed]
14. Zhu, H.; Fu, B.; Wang, S.; Zhu, L.; Zhang, L.; Jiao, L.; Wang, C. Reducing soil erosion by improving community functional diversity in semi-arid grasslands. *J. Appl. Ecol.* **2015**, *52*, 1063–1072. [CrossRef]
15. Erktan, A.; Cécillon, L.; Roose, E.; Frascaria-Lacoste, N.; Rey, F. Morphological diversity of plant barriers does not increase sediment retention in eroded marly gullies under ecological restoration. *Plant Soil* **2013**, *370*, 653–669. [CrossRef]
16. Kervroëdan, L.; Armand, R.; Saunier, M.; Faucon, M.-P. Effects of plant traits and their divergence on runoff and sediment retention in herbaceous vegetation. *Plant Soil* **2019**, *441*, 511–524. [CrossRef]
17. Guerrero-Campo, J.; Montserrat-Martí, G. Effects of soil erosion on the floristic composition of plant communities on marl in northeast Spain. *J. Veg. Sci.* **2000**, *11*, 329–336. [CrossRef]
18. Martin, C.; Pohl, M.; Alewell, C.; Körner, C.; Rixen, C. Interrill erosion at disturbed alpine sites: Effects of plant functional diversity and vegetation cover. *Basic Appl. Ecol.* **2010**, *11*, 619–626. [CrossRef]
19. Styczen, M.E.; Morgan, R.P.C. Engineering properties of vegetation. In *Slope Stabilization and Erosion Control: A Bioengineering Approach*; Morgan, R.P.C., Rickson, R.J., Eds.; E & FN SPON: London, UK, 1995; pp. 4–60, ISBN 978-1-135-83190-5.
20. Haan, C.T.; Barfield, B.J.; Hayes, J.C. *Design Hydrology and Sedimentology for Small Catchments*; Academic Press: Cambridge, MA, USA, 1994; ISBN 978-0-08-057164-5.
21. Cao, L.; Zhang, Y.; Lu, H.; Yuan, J.; Zhu, Y.; Liang, Y. Grass hedge effects on controlling soil loss from concentrated flow: A case study in the red soil region of China. *Soil Tillage Res.* **2015**, *148*, 97–105. [CrossRef]
22. Cantalice, J.R.B.; Silveira, F.P.M.; Singh, V.P.; Silva, Y.J.A.B.; Cavalcante, D.M.; Gomes, C. Interrill erosion and roughness parameters of vegetation in rangelands. *CATENA* **2017**, *148*, 111–116. [CrossRef]
23. Temple, D.M.; Robinson, K.M.; Ahring, R.M.; Davis, A.G. *Stability Design of Grass-Lined Open Channels*; Agriculture Handbook; U.S. Department of Agriculture: Washington, DC, USA, 1987.
24. Akram, S.; Yu, B.; Ghadiri, H.; Rose, C.; Hussein, J. The links between water profile, net deposition and erosion in the design and performance of stiff grass hedges. *J. Hydrol.* **2014**, *510*, 472–479. [CrossRef]
25. Burylo, M.; Rey, F.; Bochet, E.; Dutoit, T. Plant functional traits and species ability for sediment retention during concentrated flow erosion. *Plant Soil* **2012**, *353*, 135–144. [CrossRef]
26. Kervroëdan, L.; Armand, R.; Saunier, M.; Ouvry, J.-F.; Faucon, M.-P. Plant functional trait effects on runoff to design herbaceous hedges for soil erosion control. *Ecol. Eng.* **2018**, *118*, 143–151. [CrossRef]
27. Hussein, J.; Yu, B.; Ghadiri, H.; Rose, C. Prediction of surface flow hydrology and sediment retention upslope of a vetiver buffer strip. *J. Hydrol.* **2007**, *338*, 261–272. [CrossRef]
28. Cantalice, J.R.B.; Melo, R.O.; Silva, Y.J.A.B.; Cunha Filho, M.; Araújo, A.M.; Vieira, L.P.; Bezerra, S.A.; Barros, G.; Singh, V.P. Hydraulic roughness due to submerged, emergent and flexible natural vegetation in a semiarid alluvial channel. *J. Arid Environ.* **2015**, *114*, 1–7. [CrossRef]
29. Järvelä, J. Flow resistance of flexible and stiff vegetation: A flume study with natural plants. *J. Hydrol.* **2002**, *269*, 44–54. [CrossRef]

30. Pérez-Harguindeguy, N.; Díaz, S.; Garnier, E.; Lavorel, S.; Poorter, H.; Jaureguiberry, P.; Bret-Harte, M.S.; Cornwell, W.K.; Craine, J.M.; Gurvich, D.E.; et al. New handbook for standardised measurement of plant functional traits worldwide. *Aust. J. Bot.* **2013**, *61*, 167–234. [CrossRef]

31. Violle, C.; Navas, M.-L.; Vile, D.; Kazakou, E.; Fortunel, C.; Hummel, I.; Garnier, E. Let the concept of trait be functional! *Oikos* **2007**, *116*, 882–892. [CrossRef]

32. Mason, N.W.H.; Mouillot, D.; Lee, W.G.; Wilson, J.B. Functional richness, functional evenness and functional divergence: The primary components of functional diversity. *Oikos* **2005**, *111*, 112–118. [CrossRef]

33. Mason, N.W.H.; MacGillivray, K.; Steel, J.B.; Wilson, J.B. An index of functional diversity. *J. Veg. Sci.* **2003**, *14*, 571–578. [CrossRef]

34. Rao, C.R. Diversity and dissimilarity coefficients: A unified approach. *Theor. Popul. Biol.* **1982**, *21*, 24–43. [CrossRef]

35. Leps, J.; De Bello, F.; Lavorel, S.; Berman, S. Quantifying and interpreting functional diversity of natural communities: Practical considerations matter. *Preslia* **2006**, *78*, 481–501.

36. Botta-Dukát, Z. Rao's quadratic entropy as a measure of functional diversity based on multiple traits. *J. Veg. Sci.* **2005**, *16*, 533–540. [CrossRef]

37. Richet, J.-B.; Ouvry, J.-F.; Saunier, M. The role of vegetative barriers such as fascines and dense shrub hedges in catchment management to reduce runoff and erosion effects: Experimental evidence of efficiency, and conditions of use. *Ecol. Eng.* **2017**, *103*, 455–469. [CrossRef]

38. Yang, C.T. Unit stream power and sediment transport. *ASCE J. Hydraul. Div.* **1972**, *98*, 1805–1826.

39. Govers, G. Relationship between discharge, velocity and flow area for rills eroding loose, non-layered materials. *Earth Surf. Process. Landf.* **1992**, *17*, 515–528. [CrossRef]

40. Petchey, O.L.; Gaston, K.J. Functional diversity: Back to basics and looking forward. *Ecol. Lett.* **2006**, *9*, 741–758. [CrossRef]

41. Ouin, A.; Burel, F. Influence of herbaceous elements on butterfly diversity in hedgerow agricultural landscapes. *Agric. Ecosyst. Environ.* **2002**, *93*, 45–53. [CrossRef]

Article

Cover Crop Diversity as a Tool to Mitigate Vine Decline and Reduce Pathogens in Vineyard Soils

Andrew Richards [1,2,*], Mehrbod Estaki [1,3], José Ramón Úrbez-Torres [2], Pat Bowen [2], Tom Lowery [2] and Miranda Hart [1]

1   Department of Biology, University of British Columbia—Okanagan, 3187 University Way, Kelowna, BC V1V 1V7, Canada; mestaki@ucsd.edu (M.E.); miranda.hart@ubc.ca (M.H.)
2   Summerland Research and Development Centre, Agriculture and Agri-food Canada, 4200 Highway 97 South, Summerland, BC V0H 1Z0, Canada; joseramon.urbeztorres@canada.ca (J.R.Ú.-T.); pat.bowen@canada.ca (P.B.); tom.lowery@canada.ca (T.L.)
3   Department of Pediatrics, University of California San Diego, 9500 Gilman Dr, La Jolla, CA 92093, USA
*   Correspondence: andrew.richards@ubc.ca

Received: 1 March 2020; Accepted: 27 March 2020; Published: 30 March 2020

Abstract: Wine grape production is an important economic asset in many nations; however, a significant proportion of vines succumb to grapevine trunk pathogens, reducing yields and causing economic losses. Cover crops, plants that are grown in addition to main crops in order to maintain and enhance soil composition, may also serve as a line of defense against these fungal pathogens by producing volatile root exudates and/or harboring suppressive microbes. We tested whether cover crop diversity reduced disease symptoms and pathogen abundance. In two greenhouse experiments, we inoculated soil with a 10^6 conidia suspension of *Ilyonectria liriodendri*, a pathogenic fungus, then conditioned soil with cover crops for several months to investigate changes in pathogen abundance and fungal communities. After removal of cover crops, Chardonnay cuttings were grown in the same soil to assess disease symptoms. When grown alone, white mustard was the only cover crop associated with reductions in necrotic root damage and abundance of *Ilyonectria*. The suppressive effects of white mustard largely disappeared when paired with other cover crops. In this study, plant identity was more important than diversity when controlling for fungal pathogens in vineyards. This research aligns with other literature describing the suppressive potential of white mustard in vineyards.

Keywords: grapevine trunk disease; cover crops; biofumigant; young vine decline; plant-microbe interactions

---

1. Introduction

Grapevines (*Vitis vinifera* L.) are one of the most widely grown crops worldwide and an important economic commodity, especially in British Columbia where vineyards account for a total of 9652 hectares [1]. Grapevines experience multiple challenges, including competition with weeds [2], nutrient leeching [3], root lesion nematodes [4], viral infections [5], and especially fungal diseases [6] that reduce profit for growers. Although historic reports of fungal diseases exist [7], this problem has gained a considerable amount of attention in the 1990s [6] as wine grape production increased in Australia, Canada, the United States, and South Africa, among other countries [8–10].

Young vine decline (YVD) is a type of grapevine trunk disease that results in stunted growth, reduced yield, delayed fruiting, root necrosis, and eventually death in young vineyards 5–7 years old [11]. YVD occurs in British Columbia and other major wine grape regions around the world [6,12], resulting in significant economic losses [13]. YVD is considered a disease complex whereby the physical symptoms observed are a result of abiotic and biotic factors. Among many of the biotic stressors is *Ilyonectria*, a genus of soil-borne fungi and a causal agent of YVD [14]. Moreover, these fungi are generalist pathogens and are known to infect the roots of certain apple and cherry cultivars [15]. *Ilyonectria* is not only confined to vineyard soil, but also found in nurseries all over the world that often serve as breeding grounds for the pathogen [12,16,17].

In Canada, there are no commercially available fungicides or fumigants for managing young vine decline and other grapevine trunk diseases (GTDs) [18]. Methyl bromide, a once popular soil fumigant, has been phased out due to its toxicity and ozone depletion [19], and has been shown to reduce arbuscular mycorrhizal (AM) fungi and other beneficial organisms [20]. Available fumigants such as 1,3-dichloropropene and chloropicrin do not protect against the full spectrum of fungal pathogens including *Phytophthora* and *Fusarium* [21]. In addition, fungicides that are applied directly to plants as a liquid or powder coating can enter soil and accumulate overtime, reducing microbial diversity and activity [22]. Other approaches for controlling grapevine trunk pathogens include hot water treatment, in which propagation material is soaked in hot water (~50 °C) for a specified time [23,24]. This approach carries risks however, as improper procedures can damage propagation material and reduce vigor [25].

Cover cropping is a potential tool to mitigate GTDs in vineyards. Traditionally, cover crops have been grown to reduce soil erosion [26], increase available nitrogen for grapevines [27], control pests [28], and suppress weeds via allelopathy [29]. Growers also use cover crops to decrease vegetative growth in high vigor situations, which reduces canopy cover and improves the microclimates for ripening fruits [30]. Although cover crops have a long history of use in vineyards, their potential to mitigate soil-borne diseases has not been fully explored.

Existing literature highlights the biofumigant effects of brassicaceous crops (mustards/crucifers) that have exhibited suppression of soil-borne pathogens in vineyards and nurseries [31,32]. Other cover crops including forbs, legumes, and grasses may help reduce soil-borne diseases by harboring beneficial and antagonistic microbes [33,34], or via host dilution in which the risk of infection decreases with increasing host diversity [35]. The different mechanisms of suppression though these plants provide an incentive to implement cover crop diversity in vineyards as a management strategy for soil-borne diseases.

A diverse plant community can increase soil microbial diversity [36], biomass [37], and activity [38] via root exudation, rhizodeposition, and plant litter [39,40], which can improve ecosystem services. Soils from long term grasslands and forests contain plant growth-promoting rhizobacteria (PGPR), which can suppress pathogens when added to agricultural soil [41,42]. Implementation of cover crops in vineyards can increase the activity of PGPR [43,44], which are commonly found in soil [45,46]. If beneficial microbes can be isolated from nearby soil and used to reduce disease symptoms in agricultural plots, it is possible that cover crops can provide similar soil inputs and encourage proliferation of microbes that suppress fungal pathogens.

To date, most cover crop experiments use commercial rather than native plants [47,48], and the efficacy of native cover crops has not been studied extensively in vineyards [49–52]. Plant provenance may be as important as diversity due to local adaptation and coevolution between native plants and their microbial counterparts [53]. This is observed in highly specific legume-rhizobia interactions [54,55] and could hold true for other plant–microbe interactions.

In many "home vs away" studies, plants perform better when grown with soil from the same region as the plant [56–58]. Moreover, decomposition rate is increased when a plant litter is sympatric to the soil compared to allopatric soils [59,60]. Since plant–microbe interactions heavily depend on genotypic differences [61] and resource availability [40,62], microbial communities under native plants may differ compared to common cultivar cover crops, leading to differences in ecosystem services and possibly the suppression of pathogens. Given these circumstances, native cover crops may stimulate and harbor local microbial communities through more-efficient interactions based on root exudation, litter decomposition, and chemical signaling that has been subject to selective forces over many generations [63].

To evaluate the effect of cover crop diversity on GTD symptoms and the abundance of pathogenic fungi, the effects of single cover crops grown on their own were compared to the same cover crops grown together. Using native and common cover crops, we hypothesized that mixtures of cover crops would result in fewer disease symptoms, reduce pathogen abundance, and increase fungal diversity more than any plant on its own. The present study provides insight into cover crop management in vineyards, primarily in the context of disease mitigation.

2. Materials and Methods

2.1. Establishment of Experiments

In order to understand the effects of cover crop diversity on GTD symptoms, we established two separate greenhouse experiments at the Summerland Research and Development Centre (SuRDC) in Summerland, BC, Canada:

"Cultivar Study". This experiment used four cover crops that are commonly used in vineyards (Table 1). Crimson clover (*Trifolium incarnatum* L.) and buckwheat (*Fagopyrum esculentum* Moench) were purchased from a local supplier (WestCoastSeeds, Vancouver, Canada), while white mustard (*Sinapis alba* L.) and wheatgrass (*Triticum aestivum* L.) were purchased from a commercial seed supplier (Richters, Ontario, Canada).

Table 1. Selected cover crops for native study and cultivar study greenhouse experiments. Native plants were collected in the Okanagan Valley, while seeds of cultivar plants were purchased from seed suppliers. N/A = not applicable.

Treatment	Latin Binomial	Study	Diversity	Group	Life Cycle
Crimson clover	*Trifolium incarnatum*	Cultivar	1	Legume	Annual
Wheat	*Triticum aestivum*	Cultivar	1	Grass	Annual
Buckwheat	*Fagopyrum esculentum*	Cultivar	1	Forb	Annual
White mustard	*Sinapis alba*	Cultivar	1	Brassica	Annual
All cultivar		Cultivar	4		Annual
Fallow	N/A	Cultivar	0	N/A	N/A
Silky Lupine	*Lupinus sericeus*	Native	1	Legume	Perennial
Bluebunch wheatgrass	*Pseudoroegneria spicata*	Native	1	Grass	Perennial
White yarrow	*Achillea millefolium*	Native	1	Forb	Perennial
Holboell's rockcress	*Bochera hoellbelii*	Native	1	Brassica	Perennial
All native		Native	4		Perennial

"Native Study". This experiment used four plants native to the southern interior British Columbia as cover crops. We used white yarrow (*Achillea millefolium* L.) and silky lupine (*Lupinus sericeus* Pursh), which were sourced from a local supplier (Xeriscape Endemic Nursery, West Kelowna, BC, Canada) along with bluebunch wheatgrass (*Pseudoroegneria spicata* Pursh, Löve), which was collected in Summerland, BC. Holboell's rockcress (*Bochera hoellbelii* Hornem, Löve) seeds were donated by SeedsCo Community Conservation, a local native plant supplier (Table 1).

2.2. Effect of Cover Crop Diversity on Disease Symptoms

In order to observe the effect of cover crop diversity on incidence of disease, each species for the native and cultivar studies was grown on its own (monoculture) as well as with all other plants (all native or all cultivar), totaling five treatments and 10 replicates per treatment for each study (Table 1). In addition to the cover crop treatments, the cultivar study had an additional "fallow" treatment in which the soil was kept bare. This treatment was not seeded with cover crops to determine the incidence of disease and grapevine growth without the addition of inoculant or cover crop. Both experiments consisted of a randomized block design with 10 blocks (five blocks per table) to account for environmental variation inside the greenhouse and the rectangular shape of the tables. Each treatment was assigned to its block via random number generation. Treatments were standardized to four plants per pot such that monoculture pots consisted of four plants of the same species while all native and all cultivar pots consisted of one individual for each species totaling four plants (Table 1).

2.3. Location and Greenhouse Conditions

Plants were grown in a greenhouse at SuRDC (49°33′57.8″ N 119°38′10.0″ W) from 27 April 2018 to 4 March 2019. To reduce stress during warm summer months, the room was cooled by a fog system that turned on when temperatures rose above 28 °C and shade curtains were activated from 12:30 p.m. until sunset. During the spring and fall, daytime and nighttime temperatures were kept at 20 and 15 °C, respectively, with supplementary lights to maintain 15-hour days.

2.4. Soil

Soil was collected at SuRDC on 21 March 2018 from a small cherry block. This soil is described as a Skaha loamy sand which had previously harbored apples (Braeburn grafted to M.26 rootstock) until it was replanted with sweet cherry during the 2014 growing season (Table A1) [64,65]. *Fusarium, Ilyonectria,* and *Rhizoctonia* species (which are known to infect grapevine roots) were previously isolated from this site [64], increasing the likelihood of resident pathogens already in the soil. Soil was collected from the northwest guard zone, which consisted of a sweet cherry row that separated treatments from the access road. A trench (250 × 40 × 25 cm) was dug, keeping as close to the row as possible. Soil was thoroughly homogenized by hand on a large tarp and stones were removed before the soil was transferred into 3-liter nursery pots that were filled, leaving a gap of 4 cm from the top to prevent water overflow. Nursery pots were placed in SuRDC greenhouse facilities for the duration of the study.

2.5. Pathogen Incubation and Inoculation

We inoculated each pot with three isolates of *Ilyonectria liriodendri* (SuRDC 340, 60, 393) to increase the likelihood of infection. This pathogen was previously isolated from vineyards in British Columbia [6] and the isolates were selected for their ability to grow and sporulate. The addition of inoculum also ensured the presence of YVD pathogens that could infect grapevine cuttings. Single cultures of each isolate were incubated for one week (22 °C) using 5% potato dextrose agar (PDA) solution (autoclaved at 121 °C for 30 minutes). Cultures were propagated by cutting a 1 cm^2 slice of colonized agar and placing it upside down on new PDA until enough material was available for inoculation of all pots. Plates were examined under a compound light microscope to observe sporulation before inoculum preparation. A 10^6 conidia spore suspension was created for each isolate by flooding the agar plates with 1% tween solution and disturbing the surface with a metal utensil. The liquid was then passed through double-layer cheese cloth to form the stock solution. A hemocytometer was used to count conidia spores and make the specified concentration. Soil was inoculated on 24 April 2018 by pouring 45 mL of inoculum in a circle near the center of the pot.

2.6. Germination and Growth of Plants

Seeds were germinated in starter trays with an equal mixture of field soil and Sunshine Mix #4 (Sun Gro) peat/perlite mix (autoclaved at 121 °C, 1.5 hours) before transplantation into 3-liter pots on 27 April 2018. Due to the lower germination of native plants, pots were re-seeded following transplantation so that the number of plants in each pot was equal to four. Pots were watered by hand with no additional supplements and allowed to dry before subsequent watering. During the summer months, pots were watered more frequency to prevent drought and heat stress. Cultivar study plants were grown in the greenhouse until 30 July 2018, while native study plants were grown for an additional month until 3 September 2018 due to the perennial nature of the native plants. At harvest, soil was removed from roots followed by a thorough rinsing to remove as much soil as possible. Plants were bagged and taken to University of British Columbia (UBC) Okanagan where they were dried and weighed.

Vitis vinifera (Chardonnay) cuttings were collected from SuRDC on 15 February 2018 (49°33′56.2″ N 119°37′46.7″ W) and placed in a cold storage room at 2 °C until propagation. Cuttings were taken out of cold storage in June 2018 and cut into smaller pieces containing two nodes (30 cm) with a pruning tool. Canes with visible signs of mold on the surface were discarded and the remaining canes were put in a plastic container filled to a 3-cm depth of water then placed in the experimental greenhouse until the appearance of roots. Chardonnay cuttings were transplanted on 31 July 2018 (cultivar study) and 1 August 2018 (native study). Cultivar study vines were grown for approximately four and a half months while native study vines were grown for seven months to maximize exposure to pathogens. During the first week any vines that died were removed.

Initially, grapevines were given 150 mL of Miracle Gro© (20-20-20) fertilizer on a weekly basis according to manufacturer's instructions. Nutrients were reduced to (15-15-18) after six weeks followed by dilutions to induce stressful conditions (Table A2). Cultivar study grapevines were harvested on 12 December 2018. At harvest, soil was removed from roots, followed by a thorough rinsing with reverse osmosis water. Samples were placed into paper bags and held at 4 °C until January 2019. Cuttings grown in soil conditioned by native cover crop treatments were left without fertilizer from 7 December 2018 to 6 January 2019 to further induce nutrient stress. On 8 January 2019, leaves were removed from each vine to further stress the plants and increase susceptibility to pathogens. Grapevine cuttings were removed from the greenhouse on February 12 and put into cold storage for two weeks until they were destructively harvested on 4 March 2019.

2.7. Incidence of Disease

To determine the extent of necrotic tissue in Chardonnay cuttings, a cross section was cut 1 cm from the basal end of the cane and placed on a scanner (Epson Expression 1680). Images were created with Adobe Photoshop© CS2 and analyzed with WinRhizo Pro (©2013) by defining color classes representing necrotic and healthy tissue. Percent necrosis was determined by dividing the area of necrotic tissue by the total analyzed area. For native treatments an additional measurement was performed. After imagery analysis, the progression of necrosis from the basal end to the top was determined by cutting the cane into 1-cm sections and looking for signs of necrotic tissue under a dissecting microscope (VWR Bioimager BRC-1600). Disease progression was rounded to the nearest centimeter.

2.8. Molecular Data

To determine the effects of cover crop diversity and provenance on the abundance of *I. liriodendri*, we assayed the abundance of DNA extracted from soil. Soil samples were also used to measure fungal community composition and species richness. Soil samples were taken from each nursery pot after inoculation with *I. liriodendri* before seeding with cover crops (starting soil), and again before removal of cover crops (conditioned soil). Root samples were collected after four and five months of growth for the cultivar study and native study experiments, respectively. We used a digital droplet (dd) PCR assay to observe changes in the abundance of *I. liriodendri* and Illumina sequencing of the internal transcribed spacer (ITS) 2 region to uncover fungal community composition.

2.9. DNA Extraction

On 26 April 2018, three rhizosphere core samples (1 cm diameter) totaling approximately 20 g were collected from the center of each pot at a depth of five centimeters. On 30 July 2018, another set of soil samples from cultivar treatments was collected before commercial cover crops were removed using the same method described above. Soil samples from native cover crop treatments were collected on 5 September 2018 before removal of cover crops. Soil cores were homogenized and kept at −20 °C at UBC Okanagan laboratories until DNA extraction.

Soil was dried at 60 °C for 24 hours to remove water from soil, allowing a higher DNA concentration during the final elution step [66,67]. Half a gram from each sample was used for DNA isolation. DNA was extracted using the FastDNA Spin Kit for Soil (MPBio ©2018, Irvine, CA, USA) according to the manufacturer's instructions. This resulted in approximately 90 µL of eluded DNA per sample, with an average concentration of 30 ng/µL (nanodrop 1000c ©2009, Thermo Fisher Scientific, Wilmington, NC, USA). DNA was stored at −80 °C until PCR and Illumina sequencing.

After surface sterilizing, 1 gram of root subsamples was placed in a 15-mL falcon tube and frozen at −20 °C until DNA extraction. Roots were then broken down in a mortar and ground up with liquid nitrogen until very small root fragments remained. Half a gram of ground-up roots was put into lysing tubes and the rest was put back into their original falcon tubes and frozen at −20 °C. DNA extractions performed using the FastDNA Spin Kit for Soil (MPBio ©2018) with a few modifications. Lysing was performed at an intensity of 6.5 m/s instead of the standard 6.0 m/s, and initial centrifugation was extended to 10 min to promote complete separation of root tissues and nucleic acids.

2.10. Droplet Digital Assay

In order to detect the *Ilyonectria* isolates used in the inoculum, a specific primer/probe assay that targets the beta-tubulin region was designed [68]. The primer, forward 5′-CGAGGGACATACTTGTTTCCAGAG-3′ (Tm 61, GC 60%), reverse 5′-TCAACGAGGTACGCGAAATC-3′-R (Tm 62, GC 50%), and probe TGTCAAACTCACACCACGTAGGCC amplify beta-tubulin, a highly conserved region and single-copy gene, making it ideal for the quantification or spores and/or septate hyphae in soil and roots.

Reactions consisted of 10 µL Supermix (Supermix for probes no dUTP by Bio-Rad Inc., Hercules, CA, USA), 7 µL DNAse free water, 1 µL primer/probe, and 2 µL DNA, for a total volume of 20 µL. Droplets were created using the Bio-Rad QX100 Droplet Generator using the total reaction volume per sample and 70 µL of Bio-Rad Droplet Generator Oil for Probes. PCR runs were completed in the C1000 Thermal Cycler (Bio-Rad) with the following conditions: initial heating at 95 °C for 10 min, 94 °C for 1 min, and annealing at 59 °C for 2 min × 44 cycles. Fluorescence was measured using the QX 100 Droplet Reader (Bio-Rad) and Quantalife software (version 1.7.4. Bio-Rad) by selecting FAM-HEX as the fluorescence setting. The threshold was set manually at 3000 using a pure positive and environmental positive controls as a reference. For analysis, the copy number of each sample was back calculated to represent the number of copies per gram of soil and root using a formula described in Kokkoris et al. (2019) [69].

2.11. Illumina Sequencing and Bioinformatics

Illumina sequencing was completed at the Centre for Comparative Genomics and Evolutionary Bioinformatics (Dalhousie University, Halifax Nova Scotia). Amplicon sequencing of the ITS2 sub-region was performed for each treatment ($n = 5$ for cover crops, $n = 10$ for starting soil) using primers ITS86F 5′-GTGAATCATCGAATCTTTGAA-3′ and ITS4R 5′-TCCTCCGCTTATTGATATGC-3′. Samples were demultiplexed, and barcodes were removed and returned as individual per-sample fastq files from the sequencing facility.

Initial quality control and amplicon filtering was performed using the Divisive Amplicon Denoising Algorithm (DADA2 package 1.12.1) in R statistical software (R version 3.6.1, 2019) by following the DADA2 ITS Pipeline Workflow 1.9 [70]. Primers, their reverse orientation, and complements were removed from reads using cutadapt (version 2.3). Sequence reads were filtered and trimmed using filterAndTrim (DADA2) by setting standard parameters (maxN = 0, truncQ = 2, rm.phix = TRUE, and maxEE = 2). Forward and reverse reads were dereplicated using derepFastq before applying the DADA algorithm [70]. Denoising was done by pooling samples (pool = TRUE). Sharing information across samples makes it easier for singletons appearing multiple times across samples to be resolved. Paired reads were merged and an amplicon sequence variant (ASV) table was created. Finally, chimeras were removed using removeBimeraDenovo (method = "consensus") resulting in high-quality, filtered reads. The number of reads retained at each DADA2 step is shown in Table A3.

Beta diversity analyses were performed in QIIME2 (version 2019.10, https://qiime2.org) [71] and completed separately for native and cultivar studies. First, a phylogenetic tree was constructed using the q2-phylogeny plugin for QIIME2. To assign taxonomy, a reference classifier from UNITE (version 8.0) was used [72] and applied to the representative sequences from DADA2 (see above).

Native study samples were analyzed using the q2 diversity core-metrics-phylogenetic plugin. First, the ASV table containing all samples was filtered to contain only native samples. A sampling depth of 3316 was chosen based on sample B3-3 (silky lupine) because it excluded only three samples while maximizing the sampling depth. Weighted UniFrac dissimilarity [73] was used to create a distance matrix and beta diversity results were viewed via Principal Coordinates Analysis (PCoA).

Due to non-normal distribution of features and appearance of horseshoe distributions with Weighted UniFrac distance, beta diversity for cultivar study samples was performed using the DEICODE plugin (version 0.1.5) for QIIME2 [74], which creates a Robust Aitchison principal component analysis (PCA) distance matrix that handles sparse and/or non-normal datasets. A new ASV table with only cultivar samples was created. A sampling depth of 3146 was chosen, as it compromised sample exclusion and maximal sampling depth for beta diversity (see nonchim, Table A3). As with the native study, beta diversity results were viewed via PCoA.

2.12. Statistical Analyses

Data for root necrosis were transformed by taking the square root of (k-x), where k is the maximum value for percent necrotic tissue plus 1 and x is percent necrotic tissue for each sample. Disease progression of native study grapevines was normalized by taking the natural logarithm of 1+x, where x is the vertical progression of the disease, measured in centimeters.

For native study treatments, copy number per gram of root was square-root transformed to satisfy normality. After transformation, two outliers were removed from the copy number values before modelling and subsequent statistical analyses using Tukey's interquartile range (IQR). According to this method, values that are more or less than 1.5 times the IQR are removed. In the cultivar study, the copy number from root samples was cube-root transformed to meet normality assumptions. All statistical analyses were performed by fitting a linear mixed-effects model in R (R version 3.6.1, 2019, open source, http://www.r-project.org/) using the lme4 package (1.1.21). Normality was assessed using a Shapiro–Wilk normality test (stats package 3.6.1), and variance homoscedasticity was tested using Levene's test (car package 3.0.6). For each analysis, treatment was tested as a fixed factor and block as a random factor. Post hoc comparisons were completed using Tukey's honest significant difference test [75] within the emmeans package (1.4.1).

Alpha diversity of native and cultivar study samples was compared in QIIME2 with q2 diversity alpha-group-significance using Shannon evenness vectors from the q2 diversity core-metrics-phylogenetic. Overall and pairwise interactions were determined with the Kruskal–Wallis test by ranks [76] at a significance of 0.05. Beta diversity of native samples was determined via PERMANOVA in the q2 diversity beta-group-significance plugin using the Weighted UniFrac distance matrix created from the q2 diversity core-metrics-phylogenetic plugin, as it incorporates sequence abundance and phylogeny in community composition and distance between samples. Distances for native cover crop treatments were visualized using principal coordinate analysis (PCoA) with the q2-emperor plugin. For cultivar study samples, Robust Aitchison distance matrices were used from the DEICODE plugin to determine beta diversity (version 0.2.3). All PERMANOVA tests used 999 permutations and pseudo-F as the test statistic. Dispersion of native and cultivar study samples was determined with q2 diversity beta-group-significance by setting –p-method to permdisp. Dispersion tests were executed with 999 permutations and the F-value as the test statistic.

3. Results

3.1. Cover Crop Growth

Yarrow, bluebunch wheatgrass, and rockcress germinated after two to three weeks (Table A4). Silky lupine experienced lower germination rates likely due to lack of appropriate rhizobia and/or high temperatures. When grown separately, above- and below-ground biomass of bluebunch wheatgrass and white yarrow were similar to each other and significantly higher than silky lupine and rockcress. When all cover crops were grown together, above- and below-ground biomass was not different than bluebunch wheatgrass or yarrow. Cultivar monocultures varied in biomass. When grown separately, wheat yielded the highest root biomass, followed by buckwheat and clover. Crimson clover yielded the most biomass above ground followed by buckwheat then wheat. The lowest biomass measurements were observed for white mustard, in which below- and above-ground were significantly different from all other cover crop treatments. When cultivar crops were grown together, below-ground biomass was greater than all but wheat monocultures and the highest above-ground biomass.

3.2. Effect of Cover Crops on Incidence of Disease in Vines

Contrary to our hypothesis, grapevines grown in native study monocultures did not have higher rates of necrosis when compared to all plants growing together (Figure A1). Necrotic progression (evidence of necrosis from the basal to distal end) was near significant among monocultures ($p = 0.057$), with rockcress yielding the lowest average necrotic progression (Figure A2). Contrary to predictions, white mustard yielded the lowest percent necrotic tissue and was significantly different than fallow and crimson clover treatments in the cultivar study ($p = 0.035$), as seen in Figure 1. The lower necrotic damage found in grapevines growing in white mustard soil increased slightly when white mustard was grown with other cover crops.

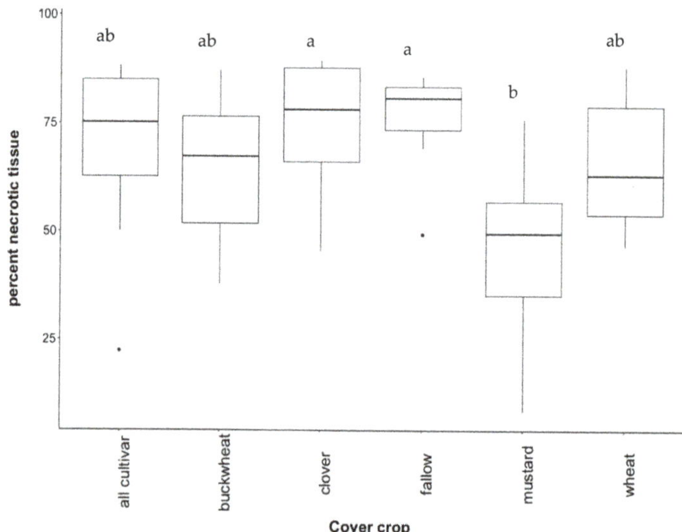

Figure 1. Percent necrotic tissue of grapevines growing in soil conditioned by cultivar cover crops. Treatments include a mixture of all plants ("all cultivar"), buckwheat ("buckwheat"), crimson clover ("clover"), uninoculated fallow ("fallow"), white mustard ("mustard"), and wheatgrass ("wheat"). Boxplots show the first and third quartile, median (middle line), range (whiskers), and circles (outliers). Letters represent statistical significance at $p < 0.05$. This section may be divided by subheadings and should provide a concise and precise depiction of the experimental results, their interpretation, and the experimental conclusions that can be drawn.

3.3. Recovery of Ilyonectria from Soil

In the native study, *I. liriodendri* was recovered from all treatments; however, its abundance varied highly between samples. Contrary to predictions, there was no significant variation between individual cover crops and when plants were grown together (Figure A3). Abundance of *I. liriodendri* was lowest in white yarrow soil while bluebunch wheatgrass yielded the highest abundance (Figure A3). In the cultivar study, abundance of *Ilyonectria* did not vary significantly between cover crop treatments (Figure 2) except for fallow, which was expected ($p < 0.001$). White mustard yielded an average of 1326 copies of *I. liriodendri* target DNA per gram of soil, the highest average copy number of all treatments, which was inconsistent with percent necrotic tissue.

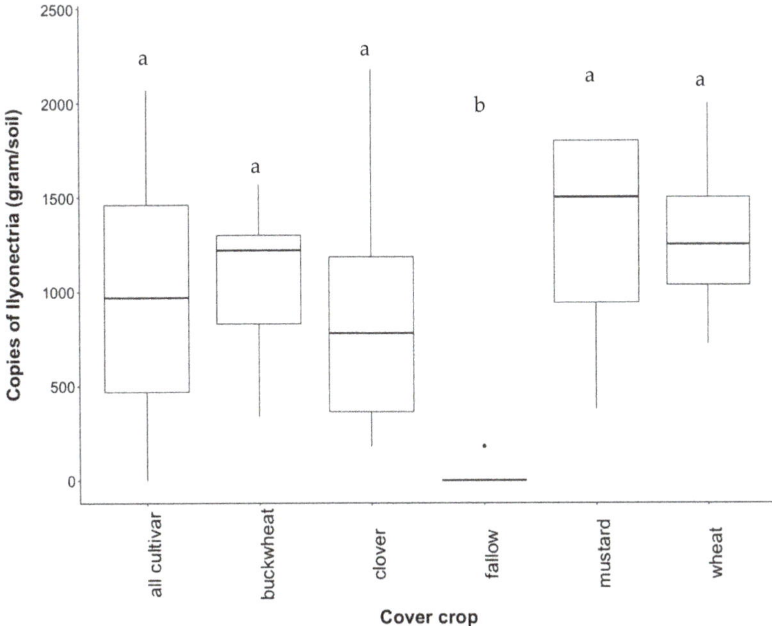

Figure 2. Recovery of *I. liriodendri* DNA from soil conditioned by cultivar cover crops. Treatments are a mixture of all plants ("all cultivar"), buckwheat ("buckwheat"), crimson clover ("clover"), uninoculated fallow ("fallow"), white mustard ("mustard"), and wheatgrass ("wheat"). Letters above boxplots represent statistical significance at $p < 0.05$.

3.4. Recovery of Ilyonectria from Roots

Contrary to predictions, *I. liriodendri* abundance did not change significantly in the native study (Figure A4). *Ilyonectria* abundance from grapevine roots was extremely variable in monocultures, with silky lupine displaying the most variability. Roots from rockcress and bluebunch wheatgrass showed the highest abundance of *I. liriodendri*, followed by white yarrow and all native (Figure A4).

Contrary to our hypothesis, abundance of *I. liriodendri* did not decrease when cultivar plants were grown together (Figure 3). Abundance of *I. liriodendri* was lowest in white mustard roots, which was consistent with the lower necrotic damage observed in grapevine cross sections from the same treatment. Abundance of *I. liriodendri* in white mustard was significantly lower compared to wheatgrass ($p = 0.041$) (Figure 3). Consistent with the digital PCR results from soil samples, roots from uninoculated fallow treatment had either zero or very small copy numbers of target DNA.

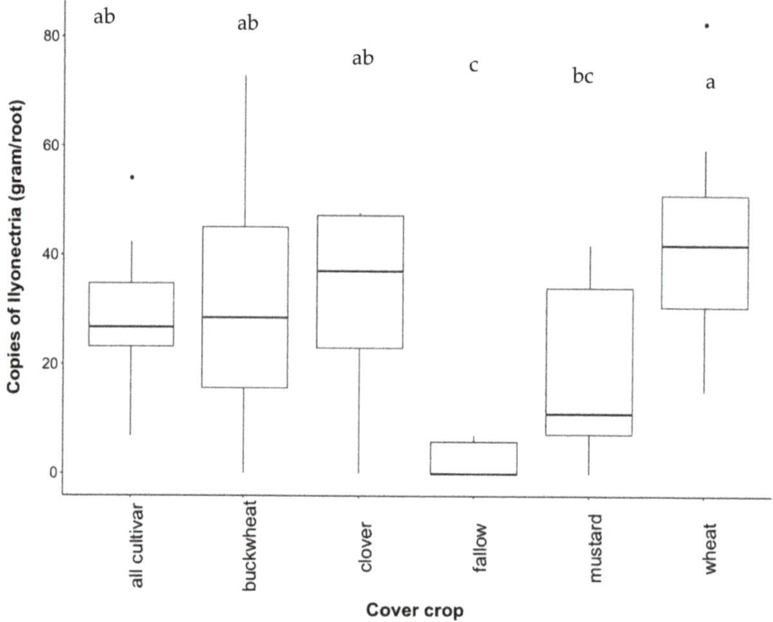

Figure 3. Abundance of *I. liriodendri* DNA from the cultivar study Chardonnay roots. The cube root of copy number was taken to normalize data. Treatments are a mixture of all plants ("all cultivar"), buckwheat ("buckwheat"), crimson clover ("clover"), uninoculated fallow ("fallow"), white mustard ("mustard"), and wheatgrass ("wheat"). Letters above boxplots represent significant differences at $p < 0.05$.

3.5. Sequence Results

A total of 2089 amplicon sequence variants (unique DNA sequences) with a combined frequency of 875,526 were present from the 111 soil samples after initial denoising and filtering. The minimum feature count per sample was 769 (bluebunch wheatgrass), while the maximum was 17,904 (soil before cover crop conditioning). The highest feature occurrence was 170,112 across all 111 samples while eight features occurred only once (0.004% of all features). A total of six phyla (one unidentified), 19 classes, 40 orders, 68 families, and 76 genera were recovered from all soil samples (Figures A5 and A6). Ascomycota yielded the highest relative frequency, followed by Basidiomycota, Mortierellomycota, and Chytridiomycota, which were present in all 111 samples. Glomeromycota was present in 78 samples, followed by an unidentified phylum that was observed in 94 samples.

3.6. Effect of Cover Crops on Fungal Diversity

3.6.1. Alpha Diversity

As predicted, alpha diversity of rhizosphere fungi increased with cover crop diversity in the native study. Silky lupine yielded the lowest fungal diversity followed by bluebunch wheatgrass. Fungal diversity was highest when all plants were grown together (Figure 4). Contrary to predictions, fungal diversity did not change with cultivar cover crops. Fungal communities were less diverse under crimson clover while buckwheat and wheatgrass were similar to the all species treatment. As expected, fallow soil contained the lowest diversity measurement, although no significant differences were detected between treatments (Figure A6).

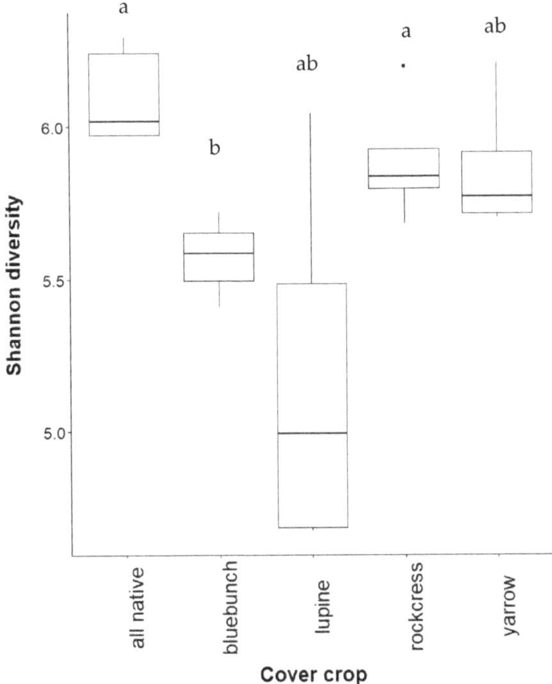

Figure 4. Shannon diversity of fungi in native study soils. Treatments are bluebunch wheatgrass ("bluebunch"), silky lupine ("lupine"), Holboell's rockcress ("rockcress"), and white yarrow ("yarrow"). Overall group significance was observed in monocultures ($p = 0.047$). Letters over treatments indicate pairwise differences at a significance level of 0.05.

3.6.2. Beta Diversity in Native and Cultivar Studies

Contrary to predictions, fungal community composition was similar among most native monocultures and when all plants were grown together ($p = 0.051$). However, community composition under bluebunch wheatgrass was distinct from white yarrow ($p = 0.036$) (Figure 5). Likewise, fungal community composition under silky lupine was different from white yarrow ($p = 0.056$). Dispersion of fungal communities (clustering) was similar under native monocultures ($p = 0.881$).

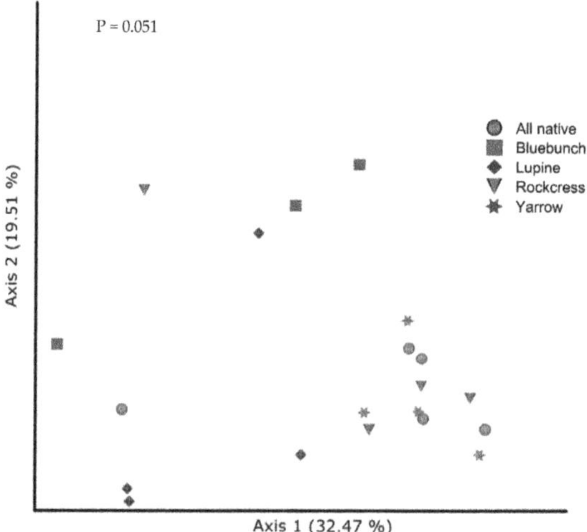

Figure 5. Principal coordinates analysis of fungal communities from native study cover crops visualized by Weighted UniFrac distance. Treatments are all species together ("All native"), bluebunch wheatgrass ("Bluebunch"), Silky lupine ("Lupine"), Holboell's rockcress ("Rockcress"), and white yarrow ("Yarrow"). Fungal communities show no significant clustering overall ($p = 0.051$); however, bluebunch wheatgrass (squares) and white yarrow (stars) reveal differences in beta diversity ($p = 0.036$, $q = 0.280$).

In the cultivar study, cover crop diversity changed community composition only in some treatments (Figure 6). All cultivar communities were distinct from fallow ($p = 0.005$) and wheatgrass ($p = 0.017$) but not others. Overall dispersion of fungal communities from monocultures was similar ($p = 0.183$).

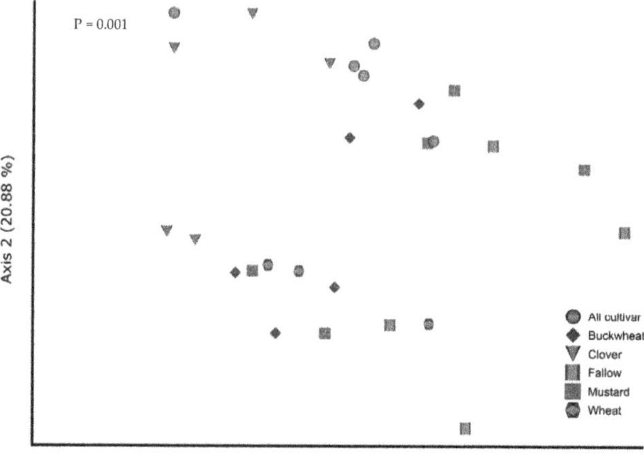

Figure 6. Principal coordinates analysis of fungal communities from cultivar cover crops visualized by Robust Aitchison distance. Treatments are all plants grown together ("All cultivar"), buckwheat ("buckwheat"), crimson clover ("Clover"), uninoculated fallow ("Fallow"), white mustard ("Mustard"), and wheatgrass ("Wheat")m with the following significant pairwise interactions: All cultivar and Fallow ($p = 0.005$, $q = 0.045$), All cultivar and wheatgrass ($p = 0.017$, $q = 0.084$), Clover and Fallow ($p = 0.006$, $q = 0.045$), clover and mustard ($p = 0.025$, $q = 0.084$), clover and wheatgrass ($p = 0.037$, $q = 0.092$).

4. Discussion

4.1. Effect of Cover Crop Diversity on Root Necrosis

Contrary to our hypothesis, cover crop diversity was not associated with necrotic root symptoms in self-rooted Chardonnay grapevines. This was true in both the native and cultivar study. One possible explanation is that the biotic properties of the soil did not change enough due to the short soil conditioning phase by cover crops. In these studies, cover crops were grown for three to four months, which translates to approximately half a growing season in the Okanagan Valley [77]. Since plant–soil feedback is not instantaneous [78], perhaps more time was needed to develop beneficial and/or antagonistic microbial communities, leading to a delay in their suppressive effects. Eisenhauer et al. (2012) [79] found that benefits from soil biota were more pronounced in long term grassland studies (four years) due to successional changes in soil microbial communities. Vogel et al. (2019) [80] further elucidated the effect of time on plant–soil feedback by showing that microbial biomass was greater in soil with a 14-year conditioning period by a specific plant community compared to new soil conditioned by the same plant community for only one year.

Another explanation could be the lack of cover crop incorporation into the soil, resulting in very little competition from other saprophytic fungi. Decomposers represent a significant group among soil microbial life and contribute to multiple ecosystem services [81]. In this experiment, we did not incorporate any litter at harvest. Instead, all cover crop material was removed including roots, which would have limited decomposer communities [82]. Most importantly, litter can contain symbiotic plant endophytes including *Trichoderma*, which are present during active plant growth and are also known to decompose litter [83,84]. The presence of *Trichoderma* could reduce GTD pathogens if they were surviving as saprophytes in plant litter, although further research is needed to determine whether stimulations of decomposer communities can reduce GTD pathogens.

In this study, the basal end of each cane was not covered with wax or another barrier, leaving a large area of vascular tissue exposed to pathogens. Such a large amount of exposed tissue in the soil would have facilitated infection even in the presence of antifungal exudates or antagonistic microbes, as below-ground wounds can serve as entry points in grapevines [85]. This is especially the case in pathogen transfer above ground in which pruning wounds left uncovered act as entry points for airborne spores [5,86].

White mustard, when grown as a monoculture, was the only cover crop that reduced necrotic tissue damage in grapevine roots. This plant matures quickly, is a high-biomass crop [87], and is known for its production of sulfur-containing glucosinolates including glucoerucin and glucoiberverin [88]. White mustard products have previously been associated with the suppression of grapevine and tree fruit pathogens [89–91]. The antifungal chemicals produced by white mustard are known to inhibit spore germination [92] and mycelial growth [93], which may have resulted in the lower incidence of necrosis observed. These results align with previous biofumigant studies that implement brassicaceous cover crops and their products [31,89,94,95].

In contrast, when white mustard was grown with other cover crops, necrotic damage was not reduced. Since each pot was standardized to four plants, only two white mustard plants grew in the soil, which would have reduced glucosinolate production. This likely reduced the concentration of antifungal compounds in the soil, allowing pathogens to proliferate more easily.

In the native study, Holboell's rockcress (a brassicaceous plant) was not associated with lower percentages of necrotic tissue. While the suppressive potential of Holboell's rockcress has not been studied in an agricultural setting, the plant matures slower due to its perennial nature, and has many natural predators including fungi [96]. Although rockcress did not show any signs of suppression in this short-term study, its persistence over multiple growing seasons and/or its degradation after maturity may contribute to the mitigation of soil-borne pathogens in vineyards.

4.2. Effect of Cover Crop Diversity on Abundance of Ilyonectria

In both studies, cover crops did not correlate with the abundance of *I. liriodendri* in the soil when grown by themselves or when grown together. This could partially be due to the absence of roots in the first centimeters of soil, where samples were taken. Overtime, the initial concentration of 1×10^6 conidia per milliliter would have diffused as pots were watered, causing spores to travel to deeper depths in the pot. Since the majority of root biomass was found below five centimeters, any effect of root exudation would have been more noticeable at lower soil depths but limited on the surface.

Consistent with percent necrotic tissue, abundance of *I. liriodendri* was lower in the white mustard monocultures. At the time of harvest, white mustard cover crops had gone to seed and had started to senesce, a period in which the metabolism of glucosinolates into antifungal isothiocyanates occurs. The breakdown and release of isothiocyanates from white mustard perhaps inhibited spore germination, reducing available inoculum during the grapevine growth stage. Antifungal compounds from brassicaceous crops can stay active for a period of 25–30 days [93,97] before they start to break down. Suppressive effects of white mustard may have been more pronounced had the plant been left to decompose in the soil [93,95,98].

In this study, *Ilyonectria* abundance in white mustard treatments was significantly lower than wheatgrass. Wheat is used in vineyards to manage soil erosion, prevent frost damage, and build organic matter [99]; however, wheat and other plants growing in a vineyard may act as off-target hosts, as has been observed in South African nurseries [48] and in Spanish vineyards [100]. In these studies, we did not examine cover crop roots for pathogens; however, it is possible that some acted as off-target hosts [48,100]. If cover crops can be colonized by *I. liriodendri* and/or other pathogens, this could maintain the spore bank and allow them to persist in soils, increasing the risk of infection. Creating a suppressive environment may require more than cover crop implementation. Changes to nutrient and watering regimes, pruning time [101], or inoculation with beneficial microbes and nearby soil may also reduce pathogens [102,103]. Indeed, there is a diverse array of fungal pathogens that infect grapevine tissues at various growth stages, which means further research is required to elucidate whether particular combinations of cover crops and pathogenic fungi can be problematic in vineyards and nurseries.

4.3. Effect of Cover Crop Diversity on Fungal Diversity

Alpha diversity of rhizosphere fungi increased with cover crop diversity in the native study but not cultivar study. The fact that microbial diversity changed under native but not cultivar cover crops perhaps implies that native plants are more dependent on resident fungi, and specifically mycorrhizal fungi, compared to plants introduced [104,105] through coevolutionary mechanisms [106]. Alternatively, carbon inputs and exudation of cultivar crops could have promoted specific fungi though positive plant–soil feedback, limiting diversity [107,108]. Since mycorrhizal fungi and resident bacteria can heavily influence functional traits—including nitrogen content, stress tolerance, morphology, leaf longevity, and pathogen resistance [109]—the presence of native plants may have stimulated these communities more than the cultivar varieties in order to maximize their fitness. Indeed, Klironomos (2003) [110] found that the frequency of positive responses from foreign plants was reduced when paired with resident AM fungi compared to the more-even distribution of responses observed when resident AM fungi and plants were paired. Alternatively, fungal diversity in the cultivar study may have been limited because the introduced plants increased the abundance of specific fungi. This has been observed in invasion studies in which the invasive plant experiences positive plant–soil feedback that allows it to outcompete native plants [107,108].

It is also possible that fungal diversity changed more under native cover crops due to a longer conditioning phase. Soil was conditioned by cultivar plants for three months whereas native plants were given four, allowing an additional month for fungi to respond to exudation, rhizodeposition [111], and root turnover [112]. In addition, root exudates and carbon deposits change as plants develop, which affects microbial communities [113]. The fact that cultivar plants matured quickly in our study perhaps led to microbial turnover whilst inputs from native plants were more consistent, allowing communities to develop overtime.

Since soil fungi are saprophytic, diversity may have increased if cover crops were left to decompose [60]. This likely would have result in compositional differences in fungal communities, as decomposers are strongly affected by plant litter type [114]. However, despite the increase in fungal diversity in cover crop mixtures and when all plants were grown together, fungal diversity was not associated with incidence of root necrosis or abundance of *I. liriodendri* in soil or roots.

Regarding pairwise interactions between treatments, alpha diversity was significantly higher under rockcress compared to bluebunch wheatgrass. Brassicaceous crops can inhibit fungal activity due to hydrolysis products of the glucosinolates they produce [115]; however, this is often limited to fungal pathogens [97], and is not widely observed in symbiotic fungi [116,117]. The fact that fungal diversity under rockcress was comparable to that of white yarrow and all plants combined suggests that rockcress did not inhibit fungi as much as other brassicaceous crops. However, alpha diversity under white mustard was also similar to other cover crops, meaning factors other than glucosinolate content contributed to alpha diversity.

4.4. Effect of Cover Crops on Community Composition

In the native study, fungal communities were dissimilar only for bluebunch wheatgrass and yarrow. Historically, white yarrow has been used as a traditional medicine in many cultures because of the phenolic compounds it produces [118] and because its extracts are known to suppress the in vitro growth of pathogenic bacteria and fungi [119]. On the contrary, bluebunch wheatgrass is not as widespread as yarrow [120], and its competitiveness is more dependent on rhizosphere microbes [121]. Given the different life history strategies employed by these plants, it is not surprising that their soil microbial communities differ.

In the cultivar study, some cover crops appeared to be more influential than others. For example, fungal communities in clover soil were distinct from those in white mustard, wheatgrass, buckwheat, and fallow soil, but not when all plants were grown together. Crimson clover produced highly branched root systems with the most above-ground biomasses out of all cover crops. Legumes are known for their mycorrhizal attributes [122], and have previously been associated with increases in fungal diversity [123], abundance of AM fungi [124], and saprophytic fungi [125]. White mustard, on the other hand, typically reduces the abundance of soil fungi relative to controls [126], although in this experiment the community composition under mustard was similar to fallow, buckwheat, wheat, and all plants grown together. At the same time, this treatment was associated with a lower incidence of root necrosis, which suggests it reduced the overall abundance of fungi associated with disease [93,97].

5. Conclusions

After a short conditioning period, we found that cover crop diversity was not associated with incidence of disease in grapevine roots. Incidence of disease was instead associated with white mustard, a common brassica cover crop. Although this apparent biofumigant effect was not observed in Holboell's rockcress (the native brassica), the results from the cultivar study align with the biofumigant literature of white mustard and other brassicaceous crops. Consistent with necrotic tissue damage, we found that white mustard was associated with a lower abundance of *I. liriodendri* in the roots of Chardonnay cuttings. However, this effect was reduced when white mustard was paired with other cover crops, and was not observed in any other monoculture.

Cover crop diversity increased fungal diversity, but only in the native study. Fungal diversity was higher in cultivar cover crops compared to fallow soil; however, there was no additive effect when all cover crops were grown together. Although not observed in this study, cover crop diversity could play a major role in the long term, especially if more diverse plant communities support diverse microbes with suppressive properties.

In summary, cover crop identity was more important than diversity for controlling fungal pathogens in grapevines. Results from the cultivar study align with other literature, which highlights the suppressive effect of brassicas. We found that when grown from seed, a brassica cover crop could offer traditional benefits such as erosion control or weed suppression as well as partially suppressing soil-borne fungi.

These results provide evidence that disease symptoms and pathogen abundance can be reduced by growing a cover crop that produces antifungal compounds. While seeding multiple cover crops confers a wide range of benefits, certain cover crops may act as vectors for fungal pathogens, thus maintaining the inoculum load. To further unveil how fungal pathogens persist in vineyard soils, future studies should focus on whether native or commercial cover crops act as vectors for GTD pathogens.

Author Contributions: Conceptualization, A.R., M.H., and J.R.Ú.-T.; methodology, A.R.; validation, P.B., T.L., and J.R.Ú.-T.; formal analysis, A.R. and M.E.; writing—original draft preparation, A.R.; writing—review and editing, M.H. and A.R.; supervision, M.H.; funding acquisition, M.H. and J.R.Ú.-T. All authors have read and agreed to the published version of the manuscript.

Funding: This research was funded by Agriculture and Agri-Food Canada, Grape Cluster activity 19, project ASC-12.

Acknowledgments: We would like to thank Xeriscape Endemic Nursery and SeedsCo Community Conservation for supplying native seeds for the experiment.

Conflicts of Interest: The authors of this paper declare no conflict of interest. The funders had no role in the design of the study; in the collection, analyses, or interpretation of data; in the writing of the manuscript; or in the decision to publish the results.

Appendix A

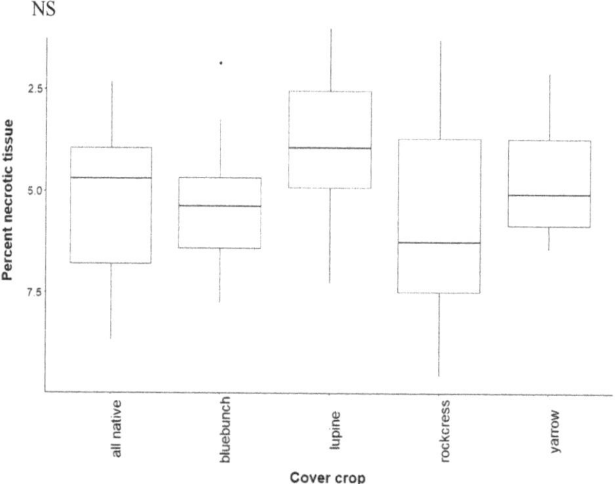

Figure A1. Percent necrotic tissue of grapevines grown in soil conditioned by native cover crops. The extent of necrosis is measured from the basal end up. Treatments are all plants grown together ("all native"), bluebunch wheatgrass ("bluebunch"), silky lupine ("lupine"), Holboell's rockcress ("rockcress"), and white yarrow ("yarrow"). Boxplots show the first and third quartile, median (middle line), range (whiskers), and circles (outliers). Data were normalized by taking the square root of the reciprocal (100.975 − x) where x is the value for percent necrosis. There was no significant difference (NS) between cover crop treatments ($p = 0.407$).

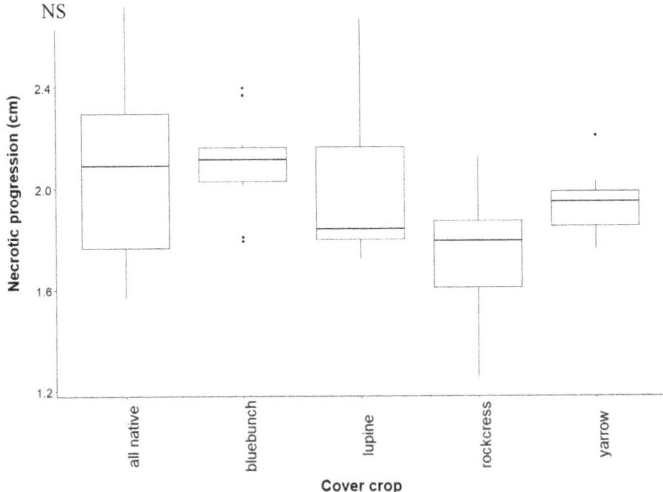

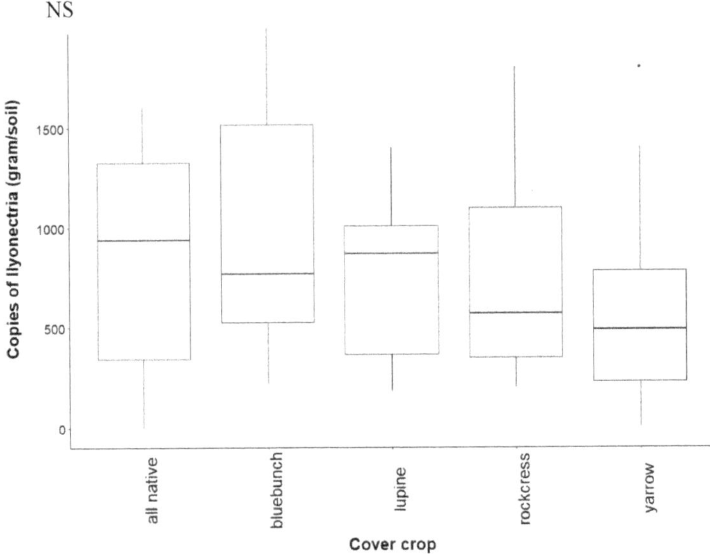

Figure A2. Progression of necrotic damage in grapevines grown in soil conditioned by native cover crops. The extent of necrosis is measured from the basal end. Treatments are all plants grown together ("all native"), bluebunch wheatgrass ("bluebunch"), silky lupine ("lupine"), Holboell's rockcress ("rockcress"), and white yarrow ("yarrow"). Boxplots show the first and third quartile, median (middle line), range (whiskers), and circles (outliers). Data were normalized by taking the natural logarithm plus 1 (log1p) of necrotic progression values. There was no significant difference (NS) between cover crop treatments.

Figure A3. Recovery of *I. liriodendri* DNA from soil after conditioning with native cover crops. Treatments are all plants grown together ("all native"), bluebunch wheatgrass ("bluebunch"), silky lupine ("lupine"), Holboell's rockcress ("rockcress"), and white yarrow ("yarrow"). Copy number did not differ significantly (NS) among treatments ($p = 0.731$).

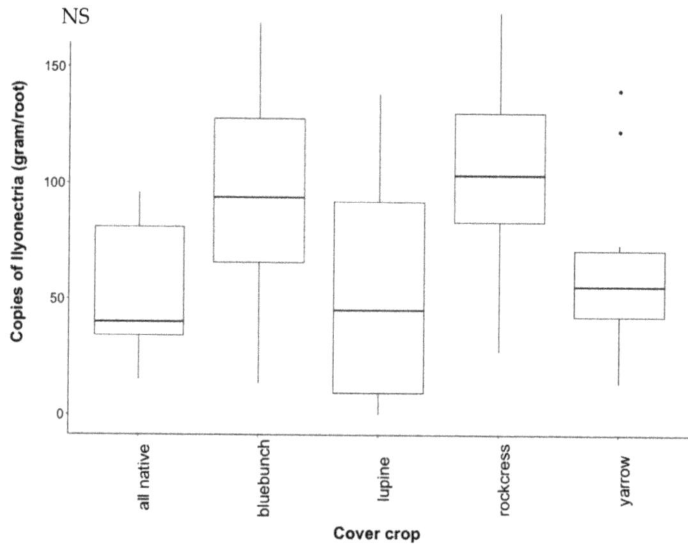

Figure A4. Recovery of *I. liriodendri* DNA from native study Chardonnay roots. Treatments are all plants grown together ("all native"), bluebunch wheatgrass ("bluebunch"), silky lupine ("lupine"), Holboell's rockcress ("rockcress"), and white yarrow ("yarrow"). Data were normalized by taking the square root of the copy number then removing outliers using Tukey's interquartile range (IQR). Copy number did not differ significantly (NS) between cover crop treatments ($p = 0.109$).

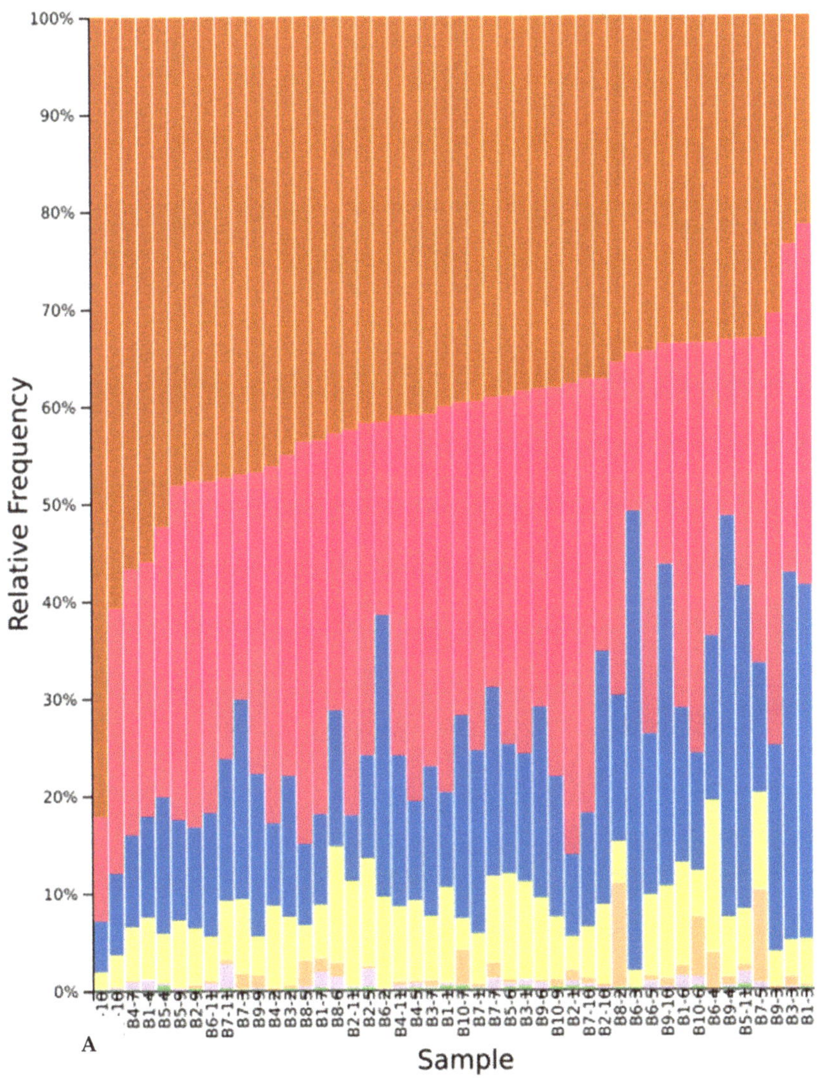

Figure A5. *Cont.*

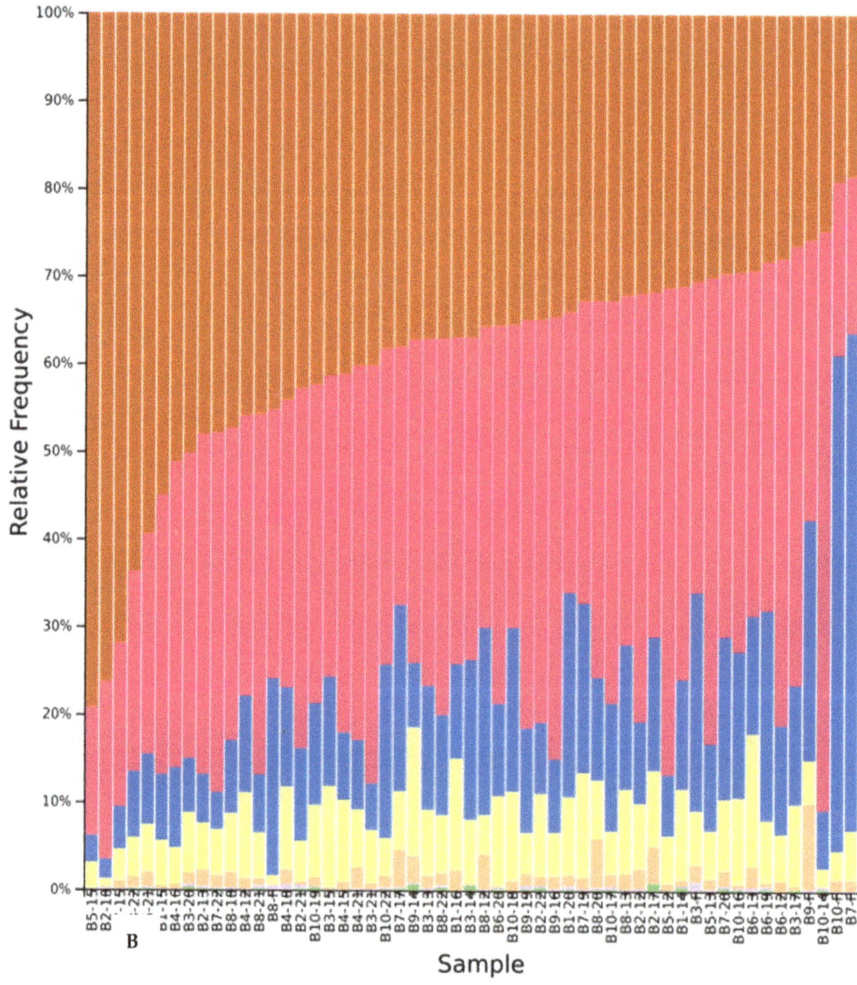

Figure A5. Relative abundance of phyla across native study samples (**A**) and cultivar study samples (**B**) after initial denoising and filtering. Samples are shown with block number followed by treatment number. Numbers 1–4 are white yarrow, Holboell's rockcress, silky lupine, and bluebunch wheatgrass, respectively while 11 is all native. Numbers 12–15 are white mustard, buckwheat, wheatgrass, and crimson clover, respectively, while 22 and 23 are all cultivar and fallow, respectively. Phyla are Ascomycota (brown), Basidiomycota (pink), Mortierellomycota (yellow), Chytridiomycota (orange), Glomeromycota (purple), and unidentified (blue and green).

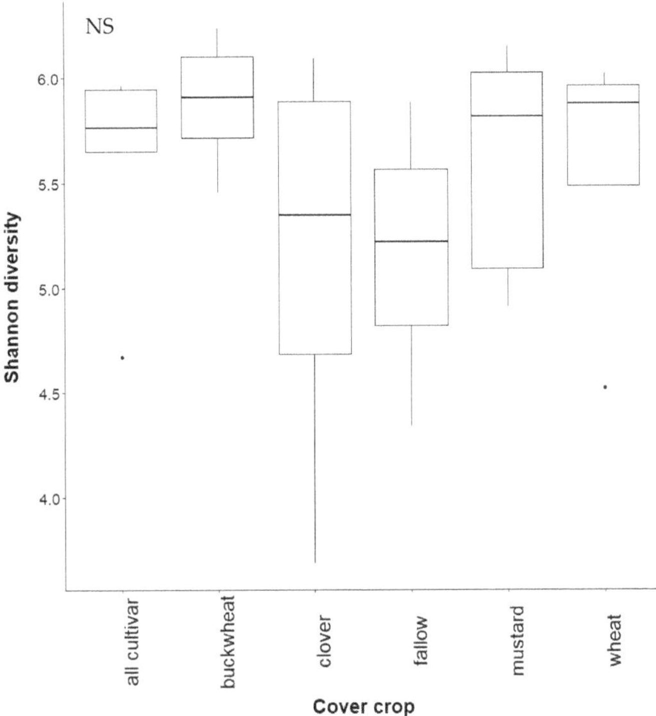

Figure A6. Shannon diversity of fungi in cultivar study soil. Treatments are all plants ("all cultivar"), buckwheat ("buckwheat"), crimson clover ("clover"), uninoculated fallow ("fallow"), white mustard ("mustard"), and wheatgrass ("wheat"). Fungal diversity did not vary significantly (NS) between treatments ($p = 0.531$).

Table A1. Site information and soil physiochemical properties adopted from Watson et al. (2018) [65].

Site Properties	Response
Fruit tree	Sweet cherry
Soil texture	Loamy sand
pH	6.9
Organic matter (%)	2.3
C/N ratio	8.5
Phosphorous (ppm)	66
Potassium (ppm)	360
Magnesium (ppm)	170
Calcium (ppm)	1330
Sodium (ppm)	25
Aluminum (ppm)	13
Sulfur (ppm)	9
Nitrate (ppm)	23

Table A2. Fertilizer application for *Vitis vinifera* cuttings in native and cultivar studies. Miracle-Gro fertilizer was used to prepare solutions of varying concentrations. Enough fertilizer was applied to cover the soil and soak through.

Date	Fertilizer Type	Amount Applied	Dilution
October 11 2018	15-15-18	150 mL	50%
October 25 2018	15-15-18	150 mL	No
November 1 2018	15-15-18	150 mL	50%
November 22 2018	15-15-18	150 mL	33%
November 29 2018	15-15-18	150 mL	40%
December 7 2018	15-15-18	150 mL	40%

Table A3. Number of reads retained at each step in the DADA2 pipeline. Samples are shown with the block number first followed by the treatment number separated by a hyphen. Numbers 1–4 are white yarrow, Holboell's rockcress, silky lupine, and bluebunch wheatgrass, respectively, while 11 is all native. Numbers 12–15 are white mustard, buckwheat, wheatgrass, and crimson clover, respectively, while 22 and 23 are all cultivar and fallow, respectively. From left to right are the initial read counts (input), reads after filtering (filtered), forward reads after denoising (denoisedF), reverse reads after denoising (denoisedR), number of reads after merging (merged), and number of reads after chimera removal (nonchim). Denoising and filtering were completed with R statistical software.

Sample	Input	filtered	DenoisedF	DenoisedR	Merged	Nonchim
B1-1	12,251	6665	6644	6644	6184	6176
B1-10	14,589	8155	8133	8127	7909	7907
B1-14	14,193	8257	8222	8183	7307	7290
B1-15	16,920	9042	9020	9013	8424	8409
B1-16	16,354	8844	8808	8789	8057	8053
B1-20	21,936	9842	9807	9799	9040	9028
B1-3	9439	4599	4589	4590	4276	4268
B1-4	13,175	6559	6524	6520	6091	6086
B1-6	10,873	6568	6514	6530	5964	5962
B1-7	10,950	6140	6113	6094	5469	5451
B10-14	6011	3446	3429	3431	3150	3146
B10-16	13,831	7968	7941	7918	7248	7233
B10-17	10,982	4927	4898	4896	4548	4506
B10-18	10,059	5622	5605	5607	5053	5029
B10-19	12,876	7437	7419	7397	6604	6579
B10-22	21,342	11,584	11,551	11,564	10,118	10,041
B10-6	29,911	15,434	15,381	15,404	14,299	14,200
B10-7	8427	4811	4787	4784	4364	4352
B10-9	13,366	6560	6530	6534	5964	5957
B10-fal	24,131	12,374	12,348	12,349	11,735	11,671
B2-1	12,412	6563	6522	6512	6117	6079
B2-10	19,272	11,002	10,973	10,965	10,166	10,161
B2-11	8626	4766	4754	4751	4461	4400
B2-12	17,980	9804	9768	9770	8801	8729
B2-13	10,050	5568	5552	5539	5178	5158
B2-17	18,591	10,111	10,080	10,067	8978	8935
B2-18	6488	3754	3734	3735	3644	3640
B2-21	20,710	10,897	10,864	10,863	10,170	10,086
B2-22	20,429	10,227	10,190	10,178	9386	9305
B2-5	12,175	7021	6998	6981	6355	6338
B2-9	15,943	8738	8696	8686	8171	8105
B3-1	14,605	8041	8015	8012	7281	7259
B3-13	8386	4782	4765	4758	4303	4296

Table A3. *Cont.*

Sample	Input	filtered	DenoisedF	DenoisedR	Merged	Nonchim
B3-14	17,549	7704	7666	7648	7118	7107
B3-15	13,922	5942	5919	5924	5398	5398
B3-17	7866	4491	4473	4469	4090	4069
B3-2	22,018	11,332	11,274	11,270	10,318	10,247
B3-20	19,794	10,473	10,436	10,427	9681	9618
B3-21	24,717	12,958	12,908	12,923	12,068	11,915
B3-22	19,204	10,218	10,182	10,191	9640	9581
B3-3	6693	3643	3631	3622	3322	3316
B3-7	20,743	10,079	10,010	10,007	9315	9237
B3-fal	18,855	10,320	10,296	10,251	9587	9503
B4-11	10,411	6031	6007	5994	5049	5049
B4-12	14,153	7003	6958	6956	6417	6403
B4-15	16,611	8708	8658	8662	7915	7840
B4-16	19,340	10,285	10,240	10,241	9650	9558
B4-18	8515	5131	5124	5122	4708	4697
B4-2	25,973	13,688	13,617	13,625	12,315	12,203
B4-21	23,882	12,563	12,498	12,473	11,638	11,530
B4-5	23,443	12,027	11,955	11,961	11,003	10,927
B4-7	20,209	10,317	10,270	10,265	9600	9531
B5-10	12,457	6726	6689	6707	6415	6378
B5-11	35,399	18,215	18,142	18,138	16,671	16,523
B5-12	24,203	13,410	13,350	13,349	12,385	12,216
B5-13	23,829	12,481	12,429	12,414	11,449	11,327
B5-15	29,538	17,066	17,035	17,022	16,486	16,416
B5-4	23,032	9630	9594	9580	8953	8943
B5-6	7570	3983	3967	3965	3565	3562
B5-9	11,398	4582	4558	4557	4269	4268
B6-11	11,204	6595	6572	6555	6071	6052
B6-12	8155	4512	4484	4491	4028	4021
B6-13	10,779	5783	5755	5744	5194	5178
B6-19	29,866	16,427	16,367	16,360	15,012	14,867
B6-2	17,688	9582	9411	9366	8683	8677
B6-20	8519	4795	4783	4782	4331	4317
B6-21	22,819	12,354	12,305	12,295	11,632	11,567
B6-3	5341	2866	2855	2851	2574	2573
B6-4	6254	847	832	835	771	769
B6-5	5377	2209	2198	2187	2000	1995
B7-1	24,124	12,565	12,508	12,504	11,635	11,547
B7-10	25,220	13,851	13,789	13,798	13,062	12,847
B7-11	9586	5557	5530	5516	5131	5111
B7-15	26,469	14,189	14,156	14,150	13,580	13,489
B7-17	16,917	8897	8856	8842	8204	8161
B7-19	13,208	8050	7998	7990	7187	7167
B7-20	10,609	6169	6151	6140	5595	5589
B7-22	29,654	16,540	16,492	16,503	15,391	15,259
B7-3	19,382	10,255	10,220	10,216	9450	9450
B7-5	11,105	5371	5353	5342	4865	4863
B7-7	12,032	6861	6824	6819	6298	6291
B7-fal	24,746	13,168	13,147	13116	12,135	12,111
B8-12	20,947	10,878	10,833	10,825	9907	9825
B8-13	14,151	7813	7771	7781	7010	7009
B8-18	13,899	6928	6888	6886	6373	6342
B8-2	26,482	14,400	14,337	14,317	13,350	13,228
B8-20	18,913	9475	9447	9451	8770	8695

Table A3. *Cont.*

Sample	Input	filtered	DenoisedF	DenoisedR	Merged	Nonchim
B8-21	14,129	7922	7901	7893	7391	7332
B8-22	18,843	9856	9801	9807	9007	8966
B8-5	20,503	10,688	10,648	10,648	10,081	10,023
B8-6	13,991	7014	6976	6985	6394	6378
B8-fal	4603	2518	2478	2466	2362	2358
B9-10	11,953	5229	5209	5202	4679	4670
B9-14	6549	3955	3938	3941	3602	3599
B9-16	11,558	6775	6752	6755	6119	6111
B9-19	11,146	5144	5116	5115	4804	4744
B9-3	10,354	4664	4648	4644	4081	4076
B9-4	13,701	6974	6945	6933	6440	6422
B9-6	9487	4444	4414	4411	4093	4089
B9-9	19,158	10,136	10,094	10,088	9287	9246
B9-fal	7037	4131	4117	4116	3850	3836
PRE-1	9227	5446	5423	5416	4963	4954
PRE-10	41,280	20,182	20,115	20,095	18,033	17,904
PRE-2	10,046	5575	5566	5559	5085	5084
PRE-3	16,164	7724	7665	7679	7081	7074
PRE-4	25,025	11,935	11,915	11,903	11,141	11,055
PRE-5	21,858	11,157	11,125	11,117	10,273	10,187
PRE-6	31,867	17,222	17,163	17,145	15,889	15,758
PRE-7	29,499	14,842	14,816	14,809	13,730	13,649
PRE-8	17,772	9465	9443	9436	8714	8689
PRE-9	22,917	12,095	12,053	12,062	11,220	11,125

Table A4. Average biomass (grams) for each cover crop treatment in native and cultivar studies. Cover crops are bluebunch wheatgrass ("bluebunch"), silky lupine ("lupine"), Holboell's rockcress ("rockcress"), white yarrow ("yarrow"), buckwheat ("buckwheat"), crimson clover ("clover"), white mustard ("mustard"), and wheatgrass ("wheat"). Letters to the right of values indicate significance at $p \leq 0.05$.

Experiment	Cover Crop Treatment	Below-Ground Biomass	Above-Ground Biomass
Native study	All native	0.885 [a]	2.715 [a]
	Bluebunch	0.795 [a]	3.041 [a]
	Lupine	0.319 [b]	0.730 [b]
	Rockcress	0.138 [b]	1.380 [b]
	Yarrow	1.056 [a]	2.822 [a]
Cultivar study	All cultivar	1.238 [a]	6.917 [a]
	Buckwheat	0.694 [b]	4.771 [b]
	Clover	0.527 [b]	7.483 [a]
	Mustard	0.121 [c]	1.834 [c]
	Wheat	2.272 [d]	3.663 [d]

References

1. Statistics Canada. 2016 Census of Agriculture. 2019. Available online: https://www.statcan.gc.ca/eng/survey/agriculture/3438 (accessed on 29 February 2020).
2. Klodd, A.E.; Eissenstat, D.M.; Wolf, T.K.; Centinari, M. "Coping with cover crop competition in mature grapevines. *Plant Soil* **2016**, *400*, 391–402. [CrossRef]
3. Messiga, A.J.; Gallant, K.S.; Sharifi, M.; Hammermeister, A.; Fuller, K.; Tango, M.; Fillmore, S. Grape yield and quality response to cover crops and amendments in a vineyard in Nova Scotia, Canada. *Am. J. Enol. Vitic.* **2016**, *67*, 77–85. [CrossRef]
4. Rahman, L.; Somers, T. Suppression of root knot nematode (Meloidogyne javanica) after incorporation of Indian mustard cv. Nemfix as green manure and seed meal in vineyards. *Australas. Plant Pathol.* **2005**, *34*, 77–83. [CrossRef]

5.   Gramaje, D.; Úrbez-Torres, J.R.; Sosnowski, M.R. Managing Grapevine Trunk Diseases With Respect to Etiology and Epidemiology: Current Strategies and Future Prospects. *Plant Dis.* **2017**, *102*, 12–39. [CrossRef]

6.   Úrbez-Torres, J.R.; Haag, P.; Bowen, P.; O'Gorman, D.T. Grapevine Trunk Diseases in British Columbia: Incidence and Characterization of the Fungal Pathogens Associated with Black Foot Disease of Grapevine. *Plant Dis.* **2013**, *98*, 456–468. [CrossRef]

7.   Maluta, D.R.; Larignon, P. Pied-noir: Mieux vaut prévenir. *Viticulture* **1991**, *11*, 71–72.

8.   Anderson, K. Lessons for Other Industries from Australia's Booming Wine Industry. *SSRN Electron. J.* **2005**. [CrossRef]

9.   Sumner, D.; Bombrun, H. An Economic Survey of the Wine and Winegrape Industry in the United States and Canada. 2001. Available online: http://www.academia.edu/download/44327963/Winegrape.pdf (accessed on 29 July 2019).

10.   Townsend, R.F.; Kirsten, J.; Vink, N. Farm size, productivity and returns to scale in agriculture revisited: A case study of wine producers in South Africa. *Agric. Econ.* **1998**, *19*, 175–180. [CrossRef]

11.   Úrbez-Torres, J.R.; Haag, P.; Bowen, P.; Lowery, T.; O'Gorman, D.T. Development of a DNA Macroarray for the Detection and Identification of Fungal Pathogens Causing Decline of Young Grapevines. *Phytopathology* **2015**, *105*, 1373–1388. [CrossRef]

12.   Gramaje, D.; di Marco, S. Identifying practices likely to have impacts on grapevine trunk disease infections: A European nursery survey. *Phytopathol. Mediterr.* **2015**, *54*, 313–324. [CrossRef]

13.   Hofstetter, V.; Buyck, B.; Croll, D.; Viret, O.; Couloux, A.; Gindro, K. What if esca disease of grapevine were not a fungal disease? *Fungal Divers.* **2012**, *54*, 51–67. [CrossRef]

14.   Morales-Cruz, A.; Figueroa-Balderas, R.; García, J.F.; Tran, E.; Rolshausen, P.E.; Baumgartner, K.; Cantu, D. Profiling grapevine trunk pathogens in planta: A case for community-targeted DNA metabarcoding. *BMC Microbiol.* **2018**, *18*, 1–14. [CrossRef] [PubMed]

15.   Watson, T.T.; Forge, T.A.; Nelson, L.M. Pseudomonads contribute to regulation of Pratylenchus penetrans (Nematoda) populations on apple. *Can. J. Bot.* **2018**, *64*, 775–785. [CrossRef] [PubMed]

16.   Álvarez-Pérez, J.M.; González-García, S.; Cobos, R.; Olego, M.Á; Ibañez, A.; Díez-Galán, A.; Garzón-Jimeno, E.; Coque, J.J.R. Use of endophytic and rhizosphere actinobacteria from grapevine plants to reduce nursery fungal graft infections that lead to young grapevine decline. *Appl. Environ. Microbiol.* **2017**, *83*, e01564–17. [CrossRef]

17.   Cardoso, M.; Diniz, I.; Cabral, A.; Rego, C.; Oliveira, H. Unveiling inoculum sources of black foot pathogens in a commercial grapevine nursery. *Phytopathol. Mediterr.* **2013**, *52*, 298–312.

18.   Canada, H. Pesticide Product Information Database. 2017. Available online: https://pesticide-registry.canada.ca/en/product-search.html (accessed on 30 January 2020).

19.   Schneider, S.M.; Rosskopf, E.N.; Leesch, J.G.; Chellemi, D.O.; Bull, C.T.; Mazzola, M. United States Department of Agriculture - Agricultural Research Service research on alternatives to methyl bromide: Pre-plant and post-harvest. *Pest Manag. Sci.* **2003**, *59*, 814–826. [CrossRef]

20.   Bendavid-Val, R.; Rabinowitch, H.D.; Katan, J.; Kapulnik, Y. Viability of VA-mycorrhizal fungi following soil solarization and fumigation. *Plant Soil* **1997**, *195*, 185–193. [CrossRef]

21.   Cabrera, J.A.; Hanson, B.D.; Gerik, J.S.; Gao, S.; Qin, R.; Wang, D. Pre-plant soil fumigation with reduced rates under low permeability films for nursery production, orchard and vineyard replanting. *Crop Prot.* **2015**, *75*, 34–39. [CrossRef]

22.   Baćmaga, M.; Wyszkowska, J.; Kucharski, J. The biochemical activity of soil contaminated with fungicides. *J. Environ. Sci. Heal.—Part B Pestic. Food Contam. Agric. Wastes* **2019**, *54*, 252–262. [CrossRef]

23.   Fourie, P.H.; Halleen, F. Proactive Control of Petri Disease of Grapevine Through Treatment of Propagation Material. *Plant Dis.* **2004**, *88*, 241–1245. [CrossRef]

24.   Gramaje, D.; Armengol, J.; Salazar, D.; López-Cortés, I.; García-Jiménez, J. Effect of hot-water treatments above 50°C on grapevine viability and survival of Petri disease pathogens. *Crop Prot.* **2009**, *28*, 280–285. [CrossRef]

25.   Waite, H.; Morton, L. Hot water treatment, trunk diseases and other critical factors in the production of high-quality grapevine planting material. *Phytopathol. Mediterr.* **2007**, *46*, 5–17.

26. Corti, G.; Cavallo, E.; Cocco, S.; Biddoccu, M.; Brecciaroli, G.; Agnelli, A. Evaluation of Erosion Intensity and Some of Its Consequences in Vineyards from Two Hilly Environments Under a Mediterranean Type of Climate, Italy. In *Soil Erosion Issues in Agriculture*; Godone, D., Stanchi, S., Eds.; IntechOpen: London, UK, 2011.

27. Bair, K.E.; Davenport, J.R.; Stevens, R.G. Release of Available Nitrogen after Incorporation of a Legume Cover Crop in Concord Grape. *HortScience* **2008**, *43*, 875–880. [CrossRef]

28. Irvin, N.A.; Hagler, J.R.; Hoddle, M.S. Measuring natural enemy dispersal from cover crops in a California vineyard. *Biol. Control* **2018**, *126*, 15–25. [CrossRef]

29. Khanh, T.D.; Chung, M.I.; Xuan, T.D.; Tawata, S. The exploitation of crop allelopathy in sustainable agricultural production. *J. Agron. Crop Sci.* **2005**, *191*, 172–184. [CrossRef]

30. Monteiro, A.; Lopes, C.M. Influence of cover crop on water use and performance of vineyard in Mediterranean Portugal. *Agric. Ecosyst. Environ.* **2007**, *121*, 336–342. [CrossRef]

31. Bleach, C.M.; Jones, E.E.; Jaspers, M.V. Biofumigation using brassicaceous plant products to control Cylindrocarpon black foot disease in New Zealand soils. *Phytopathol. Mediterr.* **2010**, *49*, 10–1007.

32. Whitelaw-Weckert, M.; Rahman, L.; Cappello, J.; Bartrop, K. Preliminary findings on the grapevine yield response to Brassica biofumigation soil treatment. *Phytopathol. Mediterr.* **2014**, *53*, 587.

33. Mazzola, M.; Granatstein, D.M.; Elfving, D.C.; Mullinix, K.; Gu, Y.H. Cultural management of microbial community structure to enhance growth of apple in replant soils. *Phytopathology* **2002**, *92*, 1363–1366. [CrossRef]

34. Wiggins, B.E.; Kinkel, L.L. Green Manures and Crop Sequences Influence Potato Diseases and Pathogen Inhibitory Activity of Indigenous Streptomycetes. *Phytopathology* **2005**, *95*, 178–185. [CrossRef] [PubMed]

35. Ostfeld, R.S.; Keesing, F. Effects of Host Diversity on Infectious Disease. *Annu. Rev. Ecol. Evol. Syst.* **2012**, *43*, 157–182. [CrossRef]

36. Steinauer, K.; Chatzinotas, A.; Eisenhauer, N. Root exudate cocktails: The link between plant diversity and soil microorganisms? *Ecol. Evol.* **2016**, *6*, 7387–7396. [CrossRef] [PubMed]

37. Eisenhauer, N.; Beßler, H.; Engels, C.; Gleixner, G.; Habekost, M.; Milcu, A.; Partsch, S.; Sabais, A.C.W.; Scherber, C.; Steinbeiss, S. Plant diversity effects on soil microorganisms support the singular hypothesis. *Ecology* **2010**, *91*, 485–496. [CrossRef] [PubMed]

38. Lange, M.; Eisenhauer, N.; Sierra, C.A.; Bessler, H.; Engels, C.; Griffiths, R.I.; Mellado-Vázquez, P.G.; Malik, A.A.; Roy, J.; Scheu, S.; et al. Plant diversity increases soil microbial activity and soil carbon storage. *Nat. Commun.* **2015**, *6*, 6707. [CrossRef]

39. Mariotte, P.; Mehrabi, Z.; Bezemer, T.M.; de Deyn, G.B.; Kulmatiski, A.; Drigo, B.; Veen, G.F.; van der Heijden, M.G.A.; Kardol, P. Plant–soil feedback: Bridging natural and agricultural sciences. *Trends Ecol. Evol.* **2018**, *33*, 129–142. [CrossRef]

40. Morgan, J.A.W.; Bending, G.D.; White, P.J. Biological costs and benefits to plant-microbe interactions in the rhizosphere. *J. Exp. Bot.* **2005**, *56*, 1729–1739. [CrossRef]

41. Garbeva, P.; Postma, J.; van Veen, J.A.; van Elsas, J.D. Effect of above-ground plant species on soil microbial community structure and its impact on suppression of Rhizoctonia solani AG3. *Environ. Microbiol.* **2006**, *8*, 233–246. [CrossRef]

42. van Elsas, J.D.; Garbeva, P.; Salles, J. Effects of agronomical measures on the microbial diversity of soils as related to the suppression of soil-borne plant pathogens. *Biodegradation* **2002**, *13*, 29–40. [CrossRef]

43. Latz, E.; Eisenhauer, N.; Rall, B.C.; Allan, E.; Roscher, C.; Scheu, S.; Jousset, A. Plant diversity improves protection against soil-borne pathogens by fostering antagonistic bacterial communities. *J. Ecol.* **2012**, *100*, 597–604. [CrossRef]

44. Latz, E.; Eisenhauer, N.; Scheu, S.; Jousset, A. Plant identity drives the expression of biocontrol factors in a rhizosphere bacterium across a plant diversity gradient. *Funct. Ecol.* **2015**, *29*, 1225–1234. [CrossRef]

45. Lee, B.D.; Dutta, S.; Ryu, H.; Yoo, S.J.; Suh, D.S.; Park, K. Induction of systemic resistance in panax ginseng against phytophthora cactorum by native bacillus amyloliquefaciens HK34. *J. Ginseng Res.* **2015**, *39*, 213–220. [CrossRef] [PubMed]

46. Shen, Z.; Ruan, Y.; Xue, C.; Zhong, S.; Li, R.; Shen, Q. Soils naturally suppressive to banana Fusarium wilt disease harbor unique bacterial communities. *Plant Soil* **2015**, *393*, 21–33. [CrossRef]

47. Irvin, N.A.; Tracy, R.P.; Thomas, M.P.; Hoddle, M. Evaluating the potential of buckwheat and cahaba vetch as nectar producing cover crops for enhancing biological control of Homalodisca vitripennis in California vineyards. *Biol. Control* **2014**, *76*, 10–18. [CrossRef]

48. Langenhoven, S.D.; Halleen, F.; Spies, C.F.J.; Stempien, E.; Mostert, L. Detection and quantification of black foot and crown and root rot pathogens in grapevine nursery soils in the Western Cape of South Africa. *Phytopathol. Mediterr.* **2018**, *57*, 519–537. [CrossRef]

49. Shields, M.W.; Tompkins, J.M.; Saville, D.J.; Meurk, C.D.; Wratten, S. Potential ecosystem service delivery by endemic plants in New Zealand vineyards: Successes and prospects. *PeerJ* **2016**, *4*, e2042. [CrossRef]

50. Tompkins, J.-M. Ecosystem Services Provided by Native New Zealand Plants in Vineyards. Ph.D. Thesis, Lincoln University, Lincoln, New Zealand, 2010.

51. Vukicevich, E.; Lowery, D.T.; Bennett, J.A.; Hart, M. Influence of Groundcover Vegetation, Soil Physicochemical Properties, and Irrigation Practices on Soil Fungi in Semi-arid Vineyards. *Front. Ecol. Evol.* **2019**, *7*, 1–10. [CrossRef]

52. Holland, T.; Vukicevich, E.; Thomsen, C.; Pogiatzis, A.; Hart, M.; Bowen, P. Arbuscular Mycorrhizal Fungi in Viticulture: Should We Use Biofertilizers? *Catal. Discov. Pract.* **2018**, *2*, 59–63. [CrossRef]

53. Revillini, D.; Gehring, C.A.; Johnson, N.C. The role of locally adapted mycorrhizas and rhizobacteria in plant–soil feedback systems. *Funct. Ecol.* **2016**, *30*, 1086–1098. [CrossRef]

54. Lupwayi, N.Z.; Kennedy, A.C. Grain legumes in Northern Great plains: Impacts on selected biological soil processes. *Agron. J.* **2007**, *99*, 1700–1709. [CrossRef]

55. Ratcliff, W.C.; Kadam, S.V.; Denison, R.F. Poly-3-hydroxybutyrate (PHB) supports survival and reproduction in starving rhizobia. *FEMS Microbiol. Ecol.* **2008**, *65*, 391–399. [CrossRef]

56. Mehrabi, Z.; Bell, T.; Lewis, O.T. Plant-soil feedbacks from 30-year family-specific soil cultures: Phylogeny, soil chemistry and plant life stage. *Ecol. Evol.* **2015**, *5*, 2333–2339. [CrossRef] [PubMed]

57. Rúa, M.A.; Antoninka, A.; Antunes, P.M.; Chaudhary, V.B.; Gehring, C.; Lamit, L.J.; Piculell, B.J.; Bever, J.D.; Zabinski, C.; Meadow, J.F.; et al. Home-field advantage? Evidence of local adaptation among plants, soil, and arbuscular mycorrhizal fungi through meta-analysis. *BMC Evol. Biol.* **2016**, *16*, 122. [CrossRef] [PubMed]

58. Smith, D.S.; Schweitzer, J.A.; Turk, P.; Bailey, J.K.; Hart, S.C.; Shuster, S.M.; Whitham, T.G. Soil-mediated local adaptation alters seedling survival and performance. *Plant Soil* **2012**, *352*, 243–251. [CrossRef]

59. Tian, K.; Kong, X.; Gao, J.; Jia, Y.; Lin, H.; He, Z.; Ji, Y.; Bei, Z.; Tian, X. Local root status: A neglected bio-factor that regulates the home-field advantage of leaf litter decomposition. *Plant Soil* **2018**, *431*, 175–189. [CrossRef]

60. Veen, G.F.; Snoek, B.L.; Bakx-Schotman, T.; Wardle, D.A.; van der Putten, W.H. Relationships between fungal community composition in decomposing leaf litter and home-field advantage effects. *Funct. Ecol.* **2019**, *33*, 1524–1535. [CrossRef]

61. Madritch, M.D.; Greene, S.L.; Lindroth, R.L. Genetic mosaics of ecosystem functioning across aspen-dominated landscapes. *Oecologia* **2009**, *160*, 119–127. [CrossRef]

62. Wei, C.; Yu, Q.; Bai, E.; Lü, X.; Li, Q.; Xia, J.; Kardol, P.; Liang, W.; Wang, Z.; Han, X.; et al. Nitrogen deposition weakens plant-microbe interactions in grassland ecosystems. *Glob. Chang. Biol.* **2013**, *19*, 3688–3697. [CrossRef]

63. Lambers, H.; Mougel, C.; Jaillard, B.; Hinsinger, P. Plant-microbe-soil interactions in the rhizosphere: An evolutionary perspective. *Plant Soil* **2009**, *321*, 83–115. [CrossRef]

64. Watson, T.; Nelson, L.; Jones, M.; Urbez-Torres, J.R. Non-Fumigant Alternative Soil Management Practices for Mitigating Replant Disease of Fruit Trees: Mechanisms Contributing to Pratylenchus Penetrans Suppression. Ph.D. Thesis, The University of British Columbia, Vancouver, BC, Canada, April 2018.

65. Watson, T.T.; Nelson, L.M.; Forge, T.A. Preplant Soil Incorporation of Compost to Mitigate Replant Disease: Soil Biological Factors Associated with Plant Growth Promotion in Orchard Soil. *Compost Sci. Util.* **2018**, *26*, 286–296. [CrossRef]

66. Overmann, J. Mahoney lake: A case study of the ecological significance of phototrophic sulfur bacteria. In *Advances in Microbial Ecology*; Jones, J.G., Ed.; Springer Science & Business Media: Berlin, Germany, 1997.

67. Lindahl, B.D.; Nilsson, R.H.; Tedersoo, L.; Abarenkov, K.; Carlsen, T.; Kjøller, R.; Kõljalg, U.; Pennanen, T.; Rosendahl, S.; Stenlid, J.; et al. Fungal community analysis by high-throughput sequencing of amplified markers—a user's guide. *New Phytol.* **2013**, *199*, 288–299. [CrossRef]

68. Holland, T.; Bowen, P.; Kokkoris, V.; Urbez-Torres, J.R.; Hart, M. Does inoculation with arbuscular mycorrhizal fungi reduce trunk disease in grapevine rootstocks? *Horticulturae* **2019**, *5*, 61. [CrossRef]

69. Kokkoris, V.; Li, Y.; Hamel, C.; Hanson, K.; Hart, M. Site specificity in establishment of a commercial arbuscular mycorrhizal fungal inoculant. *Sci. Total Environ.* **2019**, *660*, 1135–1143. [CrossRef] [PubMed]

70. Callahan, B.J.; McMurdie, P.J.; Rosen, M.J.; Han, A.W.; Johnson, A.J.A.; Holmes, S.P. DADA2: High-resolution sample inference from Illumina amplicon data. *Nat. Methods* **2016**, *13*, 581–583. [CrossRef] [PubMed]

71. Bolyen, E.; Rideout, J.R.; Dillon, M.R.; Bokulich, N.A.; Abnet, C.C.; Al-Ghalith, G.A.; Alexander, H.; Alm, E.J.; Arumugam, M.; Asnicar, F. Reproducible, interactive, scalable and extensible microbiome data science using QIIME 2. *Nat. Biotechnol.* **2019**, *37*, 852–857. [CrossRef]

72. Nilsson, R.H.; Larsson, K.H.; Taylor, A.F.S.; Bengtsson-Palme, J.; Jeppesen, T.S.; Schigel, D.; Kennedy, P.; Picard, K.; Glöckner, F.O.; Tedersoo, L.; et al. The UNITE database for molecular identification of fungi: Handling dark taxa and parallel taxonomic classifications. *Nucleic Acids Res.* **2019**, *47*, 259–264. [CrossRef]

73. Lozupone, C.; Knight, R. UniFrac: A new phylogenetic method for comparing microbial communities. *Appl. Environ. Microbiol.* **2005**, *71*, 8228–8235. [CrossRef]

74. Martino, C.; Morton, J.T.; Marotz, C.A.; Thompson, L.R.; Tripathi, A.; Knight, R.; Zengler, K. A novel sparse compositional technique reveals microbial perturbations. *Am. Soc. Microbiol.* **2019**, *4*, 1–13. [CrossRef]

75. Tukey, J.W. Comparing individual means in the analysis of variance. *Biometrics* **1949**, *5*, 99. [CrossRef]

76. Kruskal, W.H.; Wallis, W.A. Use of Ranks in One-Criterion Variance Analysis. *J. Am. Stat. Assoc.* **1952**, *47*, 583–621. [CrossRef]

77. Neilsen, D.; Smith, C.A.S.; Frank, G.; Koch, W.; Alila, Y.; Merritt, W.S.; Taylor, W.G.; Barton, M.; Hall, J.W.; Cohen, S.J.; et al. Potential impacts of climate change on water availability for crops in the Okanagan Basin, British Columbia. *Can. J. Soil. Sci* **2006**, *86*, 921–936. [CrossRef]

78. Hawkes, C.V.; Kivlin, S.N.; Du, J.; Eviner, V.T. The temporal development and additivity of plant-soil feedback in perennial grasses. *Plant Soil* **2013**, *369*, 141–150. [CrossRef]

79. Eisenhauer, N.; Reich, P.B.; Scheu, S. Increasing plant diversity effects on productivity with time due to delayed soil biota effects on plants. *Basic Appl. Ecol.* **2012**, *13*, 571–578. [CrossRef]

80. Vogel, A.; Ebeling, A.; Gleixner, G.; Roscher, C.; Scheu, S.; Ciobanu, M.; Koller-France, E.; Lange, M.; Lochner, A.; Meyer, S.T.; et al. A new experimental approach to test why biodiversity effects strengthen as ecosystems age. *Adv. Ecol. Res.* **2019**, *61*, 221–264.

81. Barrios, E. Soil biota, ecosystem services and land productivity. *Ecol. Econ.* **2007**, *64*, 269–285. [CrossRef]

82. Ball, B.A.; Bradford, M.A.; Coleman, D.C.; Hunter, M.D. Linkages between below and aboveground communities: Decomposer responses to simulated tree species loss are largely additive. *Soil Biol. Biochem.* **2009**, *41*, 1155–1163. [CrossRef]

83. Cumagun, C.J.R. Managing plant diseases and promoting sustainability and productivity with Trichoderma: The Philippine experience. *J. Agric. Sci. Technol.* **2012**, *14*, 699–714.

84. Naher, L.; Yusuf, U.K.; Ismail, A.; Hossain, K. *Trichoderma* spp.: A biocontrol agent for sustainable management of plant diseases. *Pak. J. Bot.* **2014**, *46*, 1489–1493.

85. Fourie, P.H.; Halleen, F. Occurrence of grapevine trunk disease pathogens in rootstock mother plants in South Africa. *Australas. Plant Pathol.* **2004**, *33*, 313–315. [CrossRef]

86. Cobos, R.; Mateos, R.M.; Álvarez-Pérez, J.M.; Olego, M.A.; Sevillano, S.; González-García, S.; Garzón-Jimeno, E.; Coque, J.J.R. Effectiveness of natural antifungal compounds in controlling infection by grapevine trunk disease pathogens through pruning wounds. *Appl. Environ. Microbiol.* **2015**, *81*, 6474–6483. [CrossRef]

87. Clark, A. *Managing Cover Crops Profitably*; Sustainable Agriculture Research and Education: College Park, MD, USA, 2012.

88. Fahey, J.W.; Zalcmann, A.T.; Talalay, P. The chemical diversity and distribution of glucosinolates and isothiocyanates among plants. *Phytochemistry* **2001**, *56*, 5–51. [CrossRef]

89. Berlanas, C.; Andrés-Sodupe, M.; López-Manzanares, B.; Maldonado-González, M.M.; Gramaje, D. Effect of white mustard cover crop residue, soil chemical fumigation and Trichoderma spp. root treatment on black-foot disease control in grapevine. *Pest Manag. Sci.* **2018**, *74*, 2864–2873. [CrossRef] [PubMed]

90. Mazzola, S.; Hewavitharana, M.; Strauss, S.S. Brassica seed meal soil amendments transform the rhizosphere microbiome and improve apple production through resistance to pathogen re-infestation. *Phytopathology* **2014**, *105*, 1–48. [CrossRef]

91. Wang, L.; Mazzola, M. Interaction of Brassicaceae seed meal soil amendment and apple rootstock genotype on microbiome structure and replant disease suppression. *Phytopathology* **2019**, *109*, 607–614. [CrossRef] [PubMed]

92. Drakopoulos, D.; Luz, C.; Torrijos, R.; Meca, G.; Weber, P.; Banziger, I.; Voegele, R.T.; Six, J.; Vogelgsang, S. Use of Botanicals to Suppress Different Stages of the Life Cycle of Fusarium graminearum. *Phytopathology* **2019**, *109*, 2116–2123. [CrossRef]

93. Prasad, P.; Kumar, J. Management of fusarium wilt of chickpea using brassicas as biofumigants. *Legum. Res.* **2017**, *40*, 178–182. [CrossRef]

94. Mazzola, M.; Granatstein, D.M.; Elfving, D.C.; Mullinix, K. Suppression of specific apple root pathogens by Brassica napus seed meal amendment regardless of glucosinolate content. *Phytopathology* **2001**, *91*, 673–679. [CrossRef]

95. Mowlick, S.; Yasukawa, H.; Inoue, T.; Takehara, T.; Kaku, N.; Ueki, K.; Ueki, A. Suppression of spinach wilt disease by biological soil disinfestation incorporated with Brassica juncea plants in association with changes in soil bacterial communities. *Crop Prot.* **2013**, *54*, 185–193. [CrossRef]

96. Roy, B.A. Patterns of Rust Infection as a Function of Host Genetic Diversity and Host Density in Natural Populations of the Apomictic Crucifer, Arabis holboellii. *Evolution* **1993**, *47*, 111. [CrossRef]

97. Ma, Y.; Gentry, T.; Hu, P.; Pierson, E.; Gu, M.; Yin, S. Impact of brassicaceous seed meals on the composition of the soil fungal community and the incidence of Fusarium wilt on chili pepper. *Appl. Soil Ecol.* **2015**, *90*, 41–48. [CrossRef]

98. Wen, L.; Lee-Marzano, S.; Ortiz-Ribbing, L.M.; Gruver, J.; Hartman, G.L.; Eastburn, D.M. Suppression of soilborne diseases of soybean with cover crops. *Plant Dis.* **2017**, *101*, 1918–1928. [CrossRef]

99. BC Wine Grape Council. *2010 Best Practices Guide for Grapes*; British Columbia Wine Grape Council: Peachland, BC, Canada, 2010.

100. Agustí-Brisach, C.; Gramaje, D.; León, M.; García-Jiménez, J.; Armengol, J. Evaluation of vineyard weeds as potential hosts of black-foot and petri disease pathogens. *Plant Dis.* **2011**, *95*, 803–810. [CrossRef] [PubMed]

101. Urbez-Torres, J.R.; O'Gorman, D.T.; Boule, J.; Walker, M.; Pollard-Flamand, J. Pruning time can reduce grapevine trunk diseases infection under British Columbia environmental con-ditions. *Phytopathol. Mediterr.* **2019**, *58*, 426.

102. Boruah, H.P.D.; Kumar, B.S.D. Plant Disease Suppression and Growth Promotion by a Fluorescent *Pseudomonas* Strain. *Folia Microbiol.* **2002**, *47*, 137–143. [CrossRef] [PubMed]

103. Ridout, M.; Newcombe, G. Disease suppression in winter wheat from novel symbiosis with forest fungi. *Fungal Ecol.* **2016**, *20*, 40–48. [CrossRef]

104. Inderjit; van der Putten, W. H. Impacts of soil microbial communities on exotic plant invasions. *Trends Ecol. Evol.* **2010**, *25*, 512–519. [CrossRef] [PubMed]

105. van der Putten, W.H.; Kironomos, J.N.; Wardle, D.A. Microbial ecology of biological invasions. *ISME J.* **2007**, *1*, 28–37. [CrossRef] [PubMed]

106. Gouda, S.; Kerry, R.G.; Das, G.; Paramithiotis, S.; Shin, H.S.; Patra, J.K. Revitalization of plant growth promoting rhizobacteria for sustainable development in agriculture. *Microbiol. Res.* **2018**, *206*, 131–140. [CrossRef]

107. Broz, A.K.; Manter, D.K.; Vivanco, J.M. Soil fungal abundance and diversity: Another victim of the invasive plant Centaurea maculosa. *ISME J.* **2007**, *1*, 763–765. [CrossRef]

108. Klironomos, J.N. Feedback with soil biota contributes to plant rarity and invasiveness in communities. *Nature* **2002**, *417*, 67–70. [CrossRef]

109. Friesen, M.L.; Porter, S.S.; Stark, S.C.; von Wettberg, E.J.; Sachs, J.L.; Martinez-Romero, E. Microbially Mediated Plant Functional Traits. *Annu. Rev. Ecol. Evol. Syst.* **2011**, *42*, 23–46. [CrossRef]

110. Klironomos, J.N. Variation in plant response to native and exotic arbuscular mycorrhizal fungi. *Ecology* **2003**, *84*, 2292–2301. [CrossRef]

111. Broeckling, C.D.; Broz, A.K.; Bergelson, J.; Manter, D.K.; Vivanco, J.M. Root exudates regulate soil fungal community composition and diversity. *Appl. Environ. Microbiol.* **2008**, *74*, 738–744. [CrossRef] [PubMed]

112. Rosenstock, N.; Ellström, M.; Oddsdottir, E.; Sigurdsson, B.D.; Wallander, H. Carbon sequestration and community composition of ectomycorrhizal fungi across a geothermal warming gradient in an Icelandic spruce forest. *Fungal Ecol.* **2019**, *40*, 32–42. [CrossRef]

113. Garbeva, P.; van Veen, J.A.; van Elsas, J.D. Microbial Diversity In Soil: Selection of Microbial Populations by Plant and Soil Type and Implications for Disease Suppressiveness. *Annu. Rev. Phytopathol.* **2004**, *42*, 243–270. [CrossRef]

114. Veen, G.F.; Fry, E.L.; Hooven, F.C.t.; Kardol, P.; Morriën, E.; de Long, J.R. The Role of Plant Litter in Driving Plant-Soil Feedbacks. *Front. Environ. Sci.* **2019**, *7*. [CrossRef]

115. Wang, Q.; Ma, Y.; Yang, H.; Chang, Z. Effect of biofumigation and chemical fumigation on soil microbial community structure and control of pepper Phytophthora blight. *World J. Microbiol. Biotechnol.* **2014**, *30*, 507–518. [CrossRef]

116. Glenn, M.G.; Chew, F.S.; Williams, P.H. Influence of glucosinolate content of Brassica (Cruciferae) roots on growth of vesicular–arbuscular mycorrhizal fungi. *New Phytol.* **1988**, *10*, 217–225. [CrossRef]

117. Koron, D.; Sonjak, S.; Regvar, M. Effects of non-chemical soil fumigant treatments on root colonisation with arbuscular mycorrhizal fungi and strawberry fruit production. *Crop Prot.* **2014**, *55*, 35–41. [CrossRef]

118. Vitalini, S.; Beretta, G.; Iriti, M.; Orsenigo, S.; Basilico, N.; Dall'Acqua, S.; Iorizzi, M.; Fico, G. Phenolic compounds from Achillea millefolium L. and their bioactivity. *Acta Biochim. Pol.* **2011**, *58*, 203–209. [CrossRef]

119. Candan, F.; Unlu, M.; Tepe, B.; Daferera, D.; Polissiou, M.; Sökmen, A.; Akpulat, H.A. Antioxidant and antimicrobial activity of the essential oil and methanol extracts of Achillea millefolium subsp. millefolium Afan. (Asteraceae). *J. Ethnopharmacol.* **2003**, *87*, 215–220. [CrossRef]

120. Clair, J.B.S.; Kilkenny, F.F.; Johnson, R.C.; Shaw, N.L.; Weaver, G. Genetic variation in adaptive traits and seed transfer zones for Pseudoroegneria spicata (bluebunch wheatgrass) in the northwestern United States. *Evol. Appl.* **2013**, *6*, 933–948. [CrossRef] [PubMed]

121. Callaway, R.M.; Thelen, G.C.; Barth, S.; Ramsey, P.W.; Gannon, J.E. Soil fungi alter interactions between the invader Centaurea maculosa and North American natives. *Ecology* **2004**, *85*, 062–1071. [CrossRef]

122. Hayman, D.S. Mycorrhizae of nitrogen-fixing legumes. *MIRCEN J. Appl. Microbiol. Biotechnol.* **1986**, *2*, 121–145. [CrossRef]

123. Nieva, A.S.; Bailleres, M.A.; Llames, M.E.; Taboada, M.A.; Ruiz, O.A.; Menéndez, A. Promotion of Lotus tenuis in the Flooding Pampa (Argentina) increases the soil fungal diversity. *Fungal Ecol.* **2018**, *33*, 80–91. [CrossRef]

124. Benitez, M.S.; Taheri, W.I.; Lehman, R.M. Selection of fungi by candidate cover crops. *Appl. Soil Ecol.* **2016**, *103*, 72–82. [CrossRef]

125. Sanchez, I.I.; Fultz, L.M.; Lofton, J.; Haggard, B. Soil Biological Response to Integration of Cover Crops and Nitrogen Rates in a Conservation Tillage Corn Production System. *Soil Sci. Soc. Am. J.* **2019**, *83*, 1356–1367. [CrossRef]

126. Hewavitharana, S.S.; Ruddell, D.; Mazzola, M. Carbon source-dependent antifungal and nematicidal volatiles derived during anaerobic soil disinfestation. *Eur. J. Plant Pathol.* **2014**, *140*, 39–52. [CrossRef]

MDPI
St. Alban-Anlage 66
4052 Basel
Switzerland
Tel. +41 61 683 77 34
Fax +41 61 302 89 18
www.mdpi.com

Diversity Editorial Office
E-mail: diversity@mdpi.com
www.mdpi.com/journal/diversity